# Getting a Job in Wildlife Biology:

## *What It's Like and What You Need to Know*

Stephanie Schuttler, Ph.D.

Cover design by Stephanie Schuttler using Canva®

Author photograph by Adam Schuttler

Copyright © 2020 Fancy Scientist, LLC

All rights reserved.

ISBN: 9798675375127

Library of Congress Control Number: 2020917555

Throughout this book, Dr. Stephanie Schuttler shares her experiences in wildlife biology. Your experiences and results may be different. The contents of this book and the resources provided are for informational and entertainment purposes only and do not constitute financial advice. Although Stephanie has made every effort to ensure that the information in this book was correct, the Fancy Scientist LLC. does not assume and hereby disclaims any liability to any party for any loss, damage, or disruption caused by errors or omissions, whether such errors or omissions result from negligence, accident, or any other cause.

## DEDICATION

This book is dedicated to my family. It sounds cliché, but my mom, dad, sister, brother, and husband all supported me in my love of animals and in my journey to become a wildlife biologist. As you'll find out, I did some strange things that no one in my family would ever consider doing. They never questioned this weirdness and always supported me on this unique career path. This may sound blasé to you, but in our field, it is common to have entire families that don't get it. They will make comments about "getting a real job" or not understanding or appreciating the science that we do. This was never my family. Thank you.

# CONTENTS

| | | |
|---|---|---|
| | Preface | i |
| 1 | My Unexpected Journey into Wildlife Biology | Pg 1 |
| 2 | Beginnings of a Biologist | Pg 14 |
| 3 | Leveling Up with a Ph.D. | Pg 32 |
| 4 | On the Job Market | Pg 54 |
| 5 | Know Your Destination | Pg 65 |
| 6 | Overview of Career Types | Pg 79 |
| 7 | Career Workplaces | Pg 92 |
| 8 | Decide Your Education | Pg 101 |
| 9 | Getting Valuable Experience at Any Age | Pg 110 |
| 10 | Going at it "From the Side" | Pg 126 |

# PREFACE

Several years ago, my therapist asked me during one of our sessions what my biggest problem was at that moment. I told her that it was the fact that I did not have a permanent job. At the time, I was in year two of a five-year temporary postdoc position. I had plenty of time to look. However, three years later, my biggest fear came true. I still didn't have a permanent job.

In graduate school, I sometimes worried that I was making the wrong research choices and years later, I would become pigeonholed in my career. I was told over and over again that I would not be pigeonholed; simply having a Ph.D. was an incredibly marketable skill that would land me a job. As long as I could explain how the skills that I learned from my Ph.D. could relate to the position I was applying to, I would be fine. This has not been my experience.

I have been on the job market since 2012 when I graduated with my Ph.D., and I have found it shocking how difficult it has been to land a permanent position. I have been successful at getting interviews, but there was always someone who had more experience than me or skills in an area that I was lacking for that particular job. In other words, the employers were able to pick applicants exactly suited for the position. People who applied for these jobs not only met 100% of the requirements and desired qualifications but exceeded them.

I believe the job market has changed so fast that the rules and advice that used to apply, simply don't anymore. There is also much more competition. In the past, I do believe that a Ph.D. meant that you could get lots of different types of positions at different institutions. I've even talked to high-level professionals who said they wouldn't have been competitive for the jobs that they started with in their career with the experience they had then if they were applying now.

Now that I've been in the thick of the job market for years, I would have made different choices in graduate school to have a better chance of securing the jobs I applied to. I realized that there are courses and opportunities I wish I had taken when I had the chance. But I didn't know any better at the time.

That's exactly why I wrote this book. When I began graduate school, I wish there had been such a manual. This is my attempt at writing one for you. This book is intended for anyone who needs help navigating this complicated field of wildlife biology. Although this book is written for people interested in careers in wildlife biology, a lot of the advice given will apply to those pursuing careers in conservation biology, ecology, other life and natural sciences, and probably even other sciences beyond those. It is my hope that by revealing the mistakes of my past, this book offers you some clarity in this unconventional career.

In addition to the information provided in this book, I also created an accompanying free online course with relevant resources created by myself and others such as, but not limited to blog posts, videos, examples of jobs, and other forms of commentary. I highly recommend you access and read these materials because some of them go into more depth about particular types of jobs or workplaces. It is also always important to get multiple perspectives on this career. You can find the online course full of bonus materials at https://stephanieschuttler.com/ wildlife-biology-book-bonus/.

# 1 MY UNEXPECTED JOURNEY INTO WILDLIFE BIOLOGY

I wrote this book from my personal perspective of being in the wildlife biology field for 17 years: my collective experiences, advice I've gathered from mentors, and interactions with many colleagues going through the same journey as me all at a specific point in time. Before I offer advice to you and share my insights on this atypical career, I feel that it is important for me to tell my story and how I got here myself.

Wildlife biology really is a strange vocation. Elephant researcher was not one of the careers included in the Fisher-Price® Little People® toy set I played with as a toddler. In graduate school, I can clearly remember getting my hair cut and being asked by the stylist what I did for a living. I responded that I was a wildlife biologist. Curious, she then asked me more questions. I proceeded to tell her I collected dung from forest elephants to obtain DNA for understanding their social relationships. Seriously, how does one find themselves in such a career?

When I decided I wanted to become a wildlife biologist, I had no idea what I was doing and had to learn along the way. I used every experience and interaction with a wildlife biologist, field technician, professor, graduate student, government employee, zookeeper, and conservation professional to ask questions about how they got into this career and how I could ensure my own success.

In this chapter, I provide a full overview of how I got to where I am today to lay the groundwork for the rest of this book. In the remaining chapters, I will point out in hindsight how I could have better prepared myself for getting a permanent job in wildlife biology. I think you will find it interesting how one experience or personal connection can lead to another opportunity and ultimately, change the course of your career. Let's begin.

## From Jewelry Store to Elephants

I grew up in the suburbs of Buffalo, New York, USA, in the 1980s. My family fostered my affinity for animals, especially my mom. She taught me from a young age to have compassion for all living beings, no matter how big or small, fluffy or scaly, ugly or cute. We were even lectured if we ever picked flowers when walking outside because we were ruining parts of nature that others could have enjoyed.

We are far from an outdoorsy family. I didn't even go camping until I was 20 years old in a study abroad program. Most of my interactions with wildlife occurred in my front and backyard and with common animals. My parents and I would turn over flagstones just outside our front door to look for potato bugs, put spiders we found in the corners of our ceilings into jars to look at them up close, and visit local ponds to catch frogs and snakes, which we always let go. Animals fascinated me and I felt happy in their presence.

My favorite wildlife activity took place on our annual trip to Alleghany State Park. Alleghany is the largest park in New York and located about two hours away from our home. We spent days there hiking in the forest, climbing the park's large "thunder" rocks, and walking through streams in old shoes looking for crayfish and salamanders. We called these stream walks "splash hikes."

However, none of these activities compared to my favorite: looking for animals at night using a spotlight. After dusk, we drove around the huge park, my dad holding the light, as all of us scanned the dark woods for animal movements and eye shine. The spotlight revealed a world of animals I had never seen before: raccoons, foxes, opossums, skunks, and beavers.

I went through school loving wild animals, but I never heard of any job that included this passion. The only person I knew of who had any career involving wildlife was Dr. Jane Goodall. We always subscribed to *National Geographic* magazine, and from time to time, there would be features on her living in the jungle alongside wild chimpanzees. I knew though, that I didn't have what it took to be someone like Jane Goodall.

Goodall moved from London to Gombe National Park, Tanzania in her early 20s to study chimpanzees. She lived in tents in very remote areas with more wildlife than people, and no electricity or running water. Meanwhile, my family would not even camp in Alleghany! Sleeping in a tent was completely out of the question, but they also would not stay in the rustic cabins within the state park. We always stayed in a hotel outside of the park's boundaries. I didn't think I had the courage, guts, or knowhow to do what Goodall did.

In hindsight, I actually did do something that took a lot of courage and guts at an early age. When I was around 11 years old, I received an invitation to participate in a student ambassador program to Australia and New Zealand. I knew almost nothing about these countries except that they were far away and sounded exotic. Quite honestly, I was intrigued by the images of rainforests, coral reefs, and kangaroos gracing the brochure's cover. I asked my parents for permission, applied to the program, and to my surprise, was accepted. My parents and I had never been out of the country, except for Canada, which was a 30 min drive by car from Buffalo. At the time, a lot of my family members criticized my decision to travel abroad, saying that I was too young and would not appreciate or understand the experience. Looking back though, I know this trip played a deep role in who I am today.

My parents and I went to meetings with the other students and chaperones in the program to prepare for this trip. After several months, I flew across the world with a group of people that I had only known for only a short period of time. Over three weeks, we traveled along the east coast of Australia and to northern New Zealand, visiting all types of places from big cities like Sydney to small towns full of sheep.

Before visiting, I imagined Australia as a tropical country with lush, green rainforests, but as we traveled, I realized that most of the habitat was open and arid, again, with plenty of sheep. We did hike in the rainforest, and I remember one spectacular moment where our guides showed us their sudsy hands by rubbing the leaves of a native plant between them, demonstrating how Indigenous Australians made soap and showcasing the importance of the rainforest. But I also remembered a lot of farmland and sheep. I stayed with multiple host families in cities and rural areas throughout both countries and got to experience completely new ways of life. In hindsight, I can't believe I made such a large journey so far away from home at such a young age.

I never knew what I wanted to be when I grew up or how to turn the things that I really enjoyed, wildlife and art, into viable careers. During school, I don't even remember having many career options presented to me, only the obvious ones like doctor, teacher, business owner, and lawyer. While my parents were successful and encouraged me to follow my passions, they were unable to offer me specific guidance in choosing a career. My father grew up in the projects, and one of his first jobs was in house painting. Next to the paint store, was a jewelry store. One day, a position for a stock boy opened up, he applied, and got it. Over years of hard work, he eventually became partners with the owner and then made his way to become the sole owner of our now family business. My parents didn't attend college and had no idea how to help me in pursuing my career. They assumed my high school counselors would guide me in the right direction, but they weren't much help either. We had the Internet in the late 90s when I was close to graduating, but there wasn't nearly as much information as there is today and it was difficult to navigate. As a result, I ended up applying only to universities that

I knew, which included famous Ivy league schools and local colleges.

Two universities accepted me, and one placed me on a waiting list. The first being my local state university, my "safe" option. I qualified for a scholarship at my safe school, making my tuition roughly $1,000 a semester. The other schools both cost around $30,000 a semester. I didn't want to waste money on an expensive school because I wasn't sure what career I wanted to pursue. Therefore, I decided to go to the more affordable state school nearby until I figured it all out. I thought I could always transfer to another university later if I needed to.

My university was only a fifteen-minute drive by car, so I thought it would be a waste of money to stay in the dorms there, which would cost at least thousands of dollars each semester. Secretly, I was also shy and intimidated by the other students. I ended up living at home during my college years, something that I deeply regretted for a long time afterward because I felt like I never got the "college experience." Looking back now however, I think staying home pushed me to make more adventurous decisions later on and I am grateful to not have any student loans. Living with my parents, I didn't have many distractions and it also allowed me to focus on my education.

Off I went to college and majored in biology and theater. I pursued what I thought was my passion, acting, but considering how risky and unstable this career was, I majored in biology so that I could always go to medical school as a backup career choice. My parents taught me that a doctor was a stellar choice for a career: financially secure, prestigious, and always in-demand. Biology was also one of my favorite classes in school. I even took human anatomy as an elective my senior year and was one of the brave students willing to visit a lab at a nearby college where students had dissected human cadavers for their advanced anatomy classes.

Although the human body fascinated me, I really wanted to be an actor. Or so I thought. My brother, seven years older than me, went to college for acting at the prestigious Tish School of Arts at New York University in New York City. Every Sunday night, he would call our family and tell us about all of the exciting opportunities he had. He interned for the Jim Henson Company and worked in projects involving the Muppets and *Sesame Street*, house-sat for the late Jim Henson's daughter Cheryl, went to the premier of *The Lion King* musical, was at a party where Madonna was a guest, and eventually got a job working for *Sesame Street* toys and licensing. I was so dazzled by his big city life that I wanted it for myself too.

In elementary school, I oozed creativity and was comfortable being outgoing and taking risks. I was in TAG, which stood for talented and gifted, and was my favorite class. Mr. Franko gave us fun assignments like creating new inventions, rewriting lyrics to popular songs, and designing robots. In one of my favorite assignments, my best friend and I wrote, acted in, and filmed comedy sketches similar to those seen on *Saturday Night Live* and *In*

*Living Color.*

Growing up, I loved film and television and watched all of the award shows, especially the celebrity red carpet entrances before the event even started. I acted out songs from *Les Misérables* and *Chicago* in the middle of our family room whenever my mom would leave to run errands. The couches facing inward, making it look like a small stage.

But when I moved before the summer of sixth grade, it all changed.

My new school was only twenty minutes away, but it might as well have been on the other side of the country. I didn't know anyone and had to start over. Most of the girls there had only gone to that school, and therefore, had already formed strong friendships. I bounced around between different social groups and tried to find where I fit in. Friends at my former school loved my creativity and sense of humor that I showcased in TAG, but kids at the new school just thought I was weird. They acted older than their age, and one girl who came over to my house to hang out, later made fun of me because I wanted us to play with my dog, something that my old friends loved to do. I lost the confidence I had that made me so bold in performing in front of my friends at my previous school, retracted, and became shy. Years later I found my report cards from this time period with comments from teachers remarking how shy I was and asking my parents if I was okay.

In high school, when I practiced auditioning for the school's plays and musicals in my basement, I belted out notes and emoted in the spotlight I made from focusing our overhead floodlights into one area. But when the real audition came in front of the teacher-directors, I withdrew; my movements came out small and my notes timid. Directors almost never cast me as anything but a background singer. There were a few times where I didn't even make the cut, period.

My insecurities carried over into my performances in college too. I never made any of the productions put on by our university's theater department, which is not a good sign if you are a theater major. One production involved actors to take on animal characters. In the audition, the directors asked us to improvise an animal walking across the stage. My creative mind had been so blocked up to that point, that I froze and chose the most obvious, easiest, and therefore boring choice, a wild cat of some sort. While I watched the other students turn into ostriches, turtles, and elephants, I crawled on my hands and knees, making purring noises, and flashing my teeth in a mini roar. I knew my chances were slim and I wasn't surprised that I didn't get cast.

The only positive feedback I ever remember receiving in acting class was during a scene where my character was waking up in the morning. My professor told me it was incredibly realistic and believable. Having suffered from chronic fatigue my entire life, this moment, where I was able to lay down in class, close my eyes, and rest, was genuine relief for me and mimicking waking up was close to how I really felt.

I also thought maybe theater just wasn't my thing, that I was really meant for film, which was a medium we never learned in our theater classes. Every year, I always tried to watch all of the movies up for Oscars. I enrolled in a film class and taped a list of the one hundred best movies of all time as determined by the *American Film Institute* to the wall above my desk. Almost every week, I went to the library, rented a different film, and crossed it off after watching.

I decided to get experience in film and ditched Buffalo, NY, and spent two summers in NYC with my brother. I enrolled in professional film acting classes and interned at a talent management company specializing in comedians. Being in my late teens (I have a late birthday making me young compared to others in my class), I was in an awkward age where I couldn't yet play adult roles, but I wasn't a young teen either. I enrolled in a class specifically for teens, and when classes began, I immediately felt overgrown, clearly being the oldest one. Some of the "teens" in my class didn't even look like they went through puberty. Meanwhile, I was a woman in college. To top it off, most of them had more experience than me; they were cast as leads in their school musicals, had professional dance and voice training, and some even had commercial or film experience under their belt. I felt too old to be in the room and was embarrassed, but still made it to every class.

Even though I desperately wanted to be an actor, quite honestly, at the time, I didn't have the confidence for it. My body image issues alone impaired my ability to feel comfortable in front of people. This combined with my lack of experience and difficulty in landing roles made acting never feel like a fitting career choice for me. Additionally, I only wanted to be cast in meaningful roles, where the characters or plot carried powerful messages. Auditioning for commercials in my teen class felt vapid to me.

In 2001, my brother influenced my life in an unpredictable way. In a conversation one day over a sushi dinner in the East Village, he randomly suggested I should study abroad, having loved his own experience traveling to Germany in high school. He took German for his foreign language requirement and the students in his class were given the opportunity to visit Germany for several weeks. I was envious of his trip and was inspired by him traveling so far without our parents, even though I had done it myself years before and at a younger age than him. I admired and respected my brother and just like that, I was inspired to study abroad myself.

When I returned to Buffalo for the fall semester of college, I visited the study abroad office on campus and began to search for my own adventure. I had plenty of college credits, so I was free to explore any program and any area of the world without worrying about my degree requirements. My mind was open to all sorts of possibilities.

At the office, I collected all of the program brochures related to theater, plus a few more outside of acting that simply appealed to me. When I came back home, I fanned out the brochures I had collected across my pink and white bedspread and gazed upon them. The theater programs were located in Europe, mostly Italy, England, and France, and focused on classical training like Shakespearean plays. I really wanted to keep pursuing film, so the theater programs were not very attractive to me and didn't seem helpful for my career. Going to Europe sounded lovely, but it seemed like an accessible place to travel to in the future. I could easily imagine myself vacationing there with my future husband or friends.

A few other brochures that I had gathered stood out in great contrast amongst the theater programs. Ones on marine biology displayed tropical islands surrounded by waters almost fluorescent blue on their covers. But one brochure stood out like no other one that I had collected; elephants and giraffes walked majestically across open savannas on the cover. The image matched the African landscapes I had frequently seen in nature documentaries exactly.

This brochure was for a wildlife management program in Kenya. I could never imagine myself going to Africa on my own and I questioned if I would ever have another opportunity to go in my lifetime. I was already majoring in biological sciences, so enrolling in a study abroad program for wildlife management was actually an appropriate fit for me; it wasn't as crazy as it sounded. I decided to take the plunge, applied for, and was accepted into the School for Field Studies' (SFS) program in Wildlife Management in Kenya for the summer of 2002.

A few months later, I received a packing list and started purchasing things I had never used before: Nalgene® water bottles, a Therm-a-rest® sleeping pad and sleeping bag for camping, a dorky head lamp, and wool hiking socks. My mom and I went to an outdoors store to buy the stereotypical safari shirt, but also t-shirts and zip-off field pants, among other gear. The field clothes felt so foreign to me, but I tried to get the cutest, most normal clothes I could possibly find while still meeting the standards the program had sent us. Rural Kenya was conservative; this meant that you couldn't show your knees or shoulders. We had to wear long skirts or pants, and t-shirts over shorts and tank tops. We visited the doctor where I was given many shots, cold pills containing a live virus, and a prescription for Lariam®, a malaria prophylaxis that could cause vivid dreams. After packing over-the-counter medications for all of the main categories of health symptoms, I was ready to go. I had no idea what I was in for though.

Visiting Kenya was like visiting another planet and it had certainly felt like I packed for that. It took over two days to get there: two eight-hour flights with a twelve-hour layover in London. Even though the students came from colleges across the United States, once we got to New York City, we all had the same flights to Nairobi. We met up in the JFK airport, and started to bond, especially over our short daytrip exploring London. Our travel wasn't over once we landed; we still had a four-hour drive on a combination of pothole-filled and dirt roads to our field site in southeastern Kenya.

We packed our bags into the dusty Land Rovers® and started our drive to Kilimanjaro Bush Camp (KBC), the name of our field station. I put my seatbelt on and was surprised to see it left a line of brown dirt across my lap. My brand-new pink cotton t-shirt I had bought just for this trip was already dirty and maybe even stained. I didn't realize then though that I would never be clean the entire trip, except for the few moments after my cold shower was over, and before I walked back to my room.

The next four hours felt like we were driving through a *National Geographic* spread. I had never seen anything like Kenya before; it honestly didn't seem real. As I faded in and out of wakefulness, I saw Maasai men wearing traditional Shukas, red checkered cloths draped around their bodies, while tending to their cattle and goats alongside the highway in the hot sun. The staff woke us up whenever there were zebras or giraffes that could be easily seen from the car windows. I was so excited but could only keep my eyes open for a few minutes until I fell back asleep again. At times I couldn't tell what was real and what wasn't. Sleep was so easy even though my body was tossed around in the car due to the massive potholes in the road.

When we finally got to KBC, I was exhausted and only wanted to sleep. We had a quick field site tour and I spent my first night sleeping under African skies in a banda, a small cabin-like building I shared with three other girls. For the next four weeks, I would be living with and working with 30 American students from around the country, over a dozen Kenyan staff, and four American employees. Our "campus" was small and included a main classroom/dining room, library, office, and the bandas surrounded by an electric fence to keep elephants and buffalo out. Kimana town was a couple of miles away, but we had no way of getting there besides walking. Although Kenya is an English-speaking country, Kimana was a rural town and hardly any of the locals spoke English well. Some did not even speak Kiswahili, but only their local tribal language, Maa. The town consisted of small shops selling basic supplies, eating places for locals, an outdoor marketplace for people to sell vegetables, Shukas, and tire shoes, and bars that only Kenyan men drank at.

I spent the next four weeks cut off from the rest of the world, learning about African wildlife from our Kenyan professors, our field guides, and by standing on the canvas seats of Land Rovers® in national parks. I went to Kenya really only to see and study wildlife, but from my professors I learned that wildlife and humans were inseparable issues. We had three areas of study: wildlife management, ecology, and human dimensions in wildlife. I was least interested in, and quite honestly, not really interested at all in the human dimensions portion of the program.

When it came to wildlife conservation, I saw people as the enemy. People were causing all of the problems: cutting down old-growth trees in the rainforest, dumping pollutants into rivers, and poaching beautiful species like the elephants and rhinos I was now seeing in the wild. As a young girl, I can vividly remember the news showing large fires set to stockpiles of elephant ivory in Nairobi National Park, a message from the Kenyan government that poaching would no longer be tolerated and that they were adopting a shoot-to-kill strategy.

At SFS, we learned about those controversial tactics. They worked; the poaching did largely end in Kenya, but it also created a deep division between the government and the local community. I learned that I was wrong about my interpretation of the situation, thinking that the locals were the enemy of conservation. The local community didn't hate wildlife and in fact, the pastoral Maasai coexisted peacefully amongst wildlife, grazing their cattle alongside wildebeest, buffalo, zebra, and other megafauna for hundreds of years.

Poaching occurred when people had little other opportunities to make an income or from large-scale government corruption. The local Maasai community also sometimes killed wildlife in retaliation. For example, when animals raided their crops, destroying their livelihoods in a single night with no compensation, or when they lost the ability to water their cattle in areas that they had been using for hundreds of years to recently designated protected areas. Modeled after the United States park system, the Kenyan national park rules stated that people were not allowed to enter the park's boundaries without permission, entering in a vehicle, and paying the park fees. This excluded the Maasai community near us from being able to visit the vital wetlands of Amboseli National Park, so their cattle could drink water.

Additionally, there were major changes in the government and land ownership. The Maasai were nomadic and would go from one savanna to another for new grass for their cattle to feed upon. Constantly changing where the animals foraged gave the savannas time to recover and grow new grasses before they returned again. But as people turned to agriculture over cattle ranching and larger group ranches were sold off into individual parcels, the Maasai lost their ability to range over large areas of land for grazing cattle and were forced to turn to agriculture.

Agriculture was actually less compatible with wildlife conservation in this area. The climate is arid and in order for crops to be successful, large amounts of pesticides and fertilizers have to be applied. Wildlife also lost land to graze on and often turned to farms to crop-raid furthering negative attitudes towards animals by the community. Crop-raiding animals could destroy entire fields and even kill people, but the farmers could do little in return. Often when they called the Kenya Wildlife Service to document damage and potentially receive compensation, they were slow to act, and it was not uncommon to not receive the money or enough of it to replace the lost value. Our human dimensions in wildlife professor painted these pictures vividly for us during lectures in our open-air classroom or on field trips to natural vantage points overlooking the landscape so we could see with our own eyes how people and wildlife were interwoven.

Even though I now recognized the importance of involving the local community in conservation and wildlife work, at the time, I still wasn't interested in working on this aspect of wildlife management for my career. I only wanted to study the animals themselves. Going on safaris was my favorite activity and I never tired of searching for animals across the endless horizon.

*At the park entrance to Tsavo West National Park in Kenya.*

In fact, I remember the exact moment where I decided I wanted to study wildlife as a career. We spent the day at Amboseli National Park, a small protected area approximately one hour's drive from KBC. Amboseli is famous for its picturesque views of Mount Kilimanjaro, often featured on postcards with animals in the savannas in the foreground, especially elephants. The elephants at Amboseli are incredibly easy to see and they almost seem tame because they are so used to tourism and the research that is conducted there.

Dr. Cynthia Moss is the lead scientist of that research; she is like the Jane Goodall of elephants. She lived in Kenya and has studied African savanna elephants in Amboseli for decades. She tracked the lives of every single elephant in the population from birth to death and used behavioral observations to untangle their complex social relationships.

That day, we had spent all morning on a game drive searching for animals, with the students filling out their associated assignments along the way. Going to Amboseli not only provided spectacular views of wildlife, but it was also a luxurious break from our normal field station life when we visited the lodge there. Although tourist lodges in Kenya are in the middle of remote national parks surrounded by wildlife, they have gourmet restaurants where you can order American food like hamburgers and milkshakes, drinks with ice, had swimming pools, and even air conditioning.

After we stuffed ourselves at the lunch buffet, we attended an outdoor lecture on the porch of the *Ol Tukai Lodge* led by one of Cynthia Moss' researcher assistants. The lodge was in the middle of the park and therefore surrounded by open savanna on all sides. You could count the number of trees in the distance on one hand.

The researcher explained Moss' process of observing elephants. First, all of them were individually identified by their ear tears. Kenyan savannas are dry and harsh. As a result, plants have evolved defense mechanisms like thorns to prevent animals from eating too much of their leaves and therefore killing the plant. Rose thorns are tiny in comparison to the several inches-long thorns that grow on *Acacia* trees. The needle-like thorns are sometimes shed to the ground and are extremely painful if they pierce through your sandal and into your foot (speaking from experience). As elephants forage through these spikey trees and prickly plants, and for males, when they fight, they accumulate ear tears over time, creating a unique pattern for every elephant individual. Moss and her team printed out 3 x 5 notecards with elephant ears on them, drew in every ear tear, and gave each elephant a name. They also recorded the elephants' estimated year of birth, their calves, and the family group that they belonged to.

Every day, different members of the research team drove around the park looking for elephants. When they saw elephants, they stopped the car, identified the individuals, watched them, and recorded their interactions. Cynthia Moss' fifty-year research project in Amboseli revealed savanna elephants' complex social lives, behaviors that suggest elephants have emotions, and their keen intelligence. For instance, Moss' research determined that elephants can not only tell the difference between tourists and Kenyans, but also Kenyans that belong to different tribes. It is one of the longest running continuous animal field studies.

While I was soaking in every word from the researcher like a sponge, another student noticed an elephant grazing behind her far in the savanna background. We were approaching the question portion of the lecture, and this student raised their hand to test the researcher's identification skills she had just finished explaining to us. The elephant was far in the distance, and even slightly obscured from the haze created by the heat of the day. The scientist picked up her binoculars, squinted, and stared for a few moments into the lens. She then put down the binoculars and stated the elephant's name confidently. I was hooked. I thought this was the coolest job in the world, and I wanted to conduct research on the conservation of remarkable species like elephants.

Ironically, I traveled all the way to Africa to discover my new career choice, something that I never thought I could do years ago when reading about Jane Goodall. However, even though our campus is located in Africa, being a student at SFS was nothing like Goodall's experiences in the field. We had running water, electricity, showers (although though they were cold), and a chef who cooked lunch and dinners for us. Students complained about missing pizza and ice cream, but our food was high luxury compared to the typical fieldwork rations of sardines, rice, and pasta.

Despite being in a remote location, we were still a large group at about 30 students and 20 staff members. This meant a lot of people to talk to and play games with for entertainment since we didn't have television or Internet. Because we all essentially lived together for a month, I developed close relationships with both the students and staff, started overcoming my shyness, and becoming more of my true self: funny, silly, and creative.

At SFS, I learned there was a lot of variation in careers in wildlife biology, that I could study animals without having to live for years in a tent. My experiences in Kenya triggered my memories of looking for wildlife in Alleghany State Park as a child and served as confirmation that this was the perfect career for me. I had found my calling, and as our study abroad session came to a close, I was excited to apply this newfound knowledge back to my studies in the United States.

When I got back to Buffalo, I knew I had to do a few things right away to get my new career trajectory on track. When the fall semester of my senior year began, I dropped theater down to a minor and kept biological sciences as my sole major. After getting advice from the professors in Kenya, I learned that I needed research experience and that the best way to get that was by volunteering in a professor's lab at my university. I looked up every biology professors' research on the department website and emailed any that were conducting research related to wildlife. We had no wildlife biology department and finding professors' research that was relevant was a stretch as most focused on molecular, cellular, or developmental biology. Nevertheless, I did the best I could for what I had to work with.

I emailed my ecology professor and asked if I could volunteer in her lab on researching the beneficial relationships between coral and their algal symbionts. She was not able to offer me a position but referred me to another professor conducting behavioral research that was accepting students. Given my love for the elephant behavior research I had learned about in Amboseli, this seemed like an even better fit. But my new study species shared little in common with elephants.

I volunteered in a lab that researched sexual selection in flour beetles. Flour beetles were nowhere near the animals I was interested in studying later on in my career, and this research had zero conservation implications. In fact, they are a pest species for agriculture, but it was a start. I don't remember the exact research questions I was helping to answer, but the work involved female mate choice and underlying genetic factors.

The experiments consisted of putting beetles into small vials of flour, letting weeks pass by to give them time to mate, and eventually crushing the beetles to extract their DNA. As an animal lover, I hated crushing the beetles even though they were just only beetles. Luckily, my role almost exclusively dealt with the extracted DNA. In the lab, I created "cocktails" or tiny mixtures of various reagents along with the DNA extract in a small tube. I placed this tube in a machine that would vary temperatures so that the DNA could be separated and replicated, a technique called Polymerase Chain Reactions (PCR). After the DNA amplified, I ran the products from the tube using gel electrophoresis to obtain results. For this procedure, I made and melted gels, a substance similar to, but firmer than Jell-O®, put them in a machine under a solution, and used electricity to pull apart the DNA. Shorter fragments of DNA run through the gel faster than bigger ones. After you let the DNA separate, you take the gel out and look at the results under UV light.

During my entire time volunteering in this lab, I pretty much only did PCRs and gels. I was happy to get some experience under my belt, but the work was honestly boring and repetitive. I remember reading several scientific papers related to the research, but I was mostly disconnected mentally and physically from the research questions. The gels were located in a separate room across the hall from the lab, and I only interacted with the postdoc leading the project. I never knew the professor, graduate students, or other undergraduate students (if there were any) in the lab. This was in huge contrast to my experience at SFS, where we not only bonded by conducting research in student groups, but also ate and socialized together. We were intricately involved in the research and it was a central part of our daily discussions.

During my final year, I made sure to pack my remaining semesters with courses related to wildlife biology. But again, at my university, this was a stretch. Courses like mammalogy, ornithology, and animal behavior were in the course catalog, but they were never offered. Apparently, those professors retired. I was able to enroll in a tropical marine ecology course and volunteered as an undergraduate teaching assistant for the course general ecology, which provided me with a little bit more experience to put on my resume before I finally graduated in 2003.

# 2 BEGINNINGS OF A BIOLOGIST

After I graduated from college, I knew I wanted to take some time off from school to get hands-on experience and a better understanding of what research was like. My sister warned me that once I was away from school, I wouldn't want to go back, but I knew that I would go to graduate school and probably pursue my Ph.D. I just had no idea what type of research I wanted to do. My epiphany in Kenya happened so late in my college experience; it gave me little time to really think about what I was interested in or to try things out.

My first internship was the hardest to get. I studied monster.com, a website including career advice, as hard as I had studied for my final exams in college. It was hard to apply their examples, which all were from the business world to wildlife biology, but I learned everything I could about writing a compelling cover letter and using action-oriented verbs in my resume. However, this didn't make up for my lack of experience.

With no field experience especially, I was outright rejected from all different types of internships and temporary research positions (e.g., field assistants and technicians). Although we did do some research in Kenya, I was only there for a month, and also enrolled in courses at the same time. It wasn't enough time to demonstrate any significant time spent in the field. Having loved my experience in Kenya so much, I applied enthusiastically for the year-long internship with SFS there but was flat out rejected. After dozens of rejections, I finally had some luck and landed a position with a new program created by the Chicago Botanic Garden in collaboration with the US Department of Interior within the federal government. The internships took place in government field offices within the National Park Service or Bureau of Land Management departments all over the western United States.

Me and the dozens of students selected for the program met up for a training session in Chicago, IL at the beginning of summer. I met students from all over the country and with varying levels of field experience. Just like SFS, we learned skills together for our upcoming jobs in a shared classroom, ate meals together outside in the botanic garden, and even spent a night eating deep-dish pizza and dancing in downtown Chicago. But these moments of bonding as a group were short-lived. After a few days, we all dispersed to our various outposts in remote parts of the western US.

The program sent me to the Bureau of Land Management's (BLM) Arizona Strip Field Office in St. George, Utah. I was so excited to get stationed in Arizona and have my first real field experience. I envisioned living in a big, cool city like Phoenix, surrounded by mountains and saguaro cactus. But living in Arizona wasn't an option. Our fieldwork took place in northwestern Arizona, but our office was located a few miles north of the Arizona-Utah border. Northwestern Arizona is mostly made up of unoccupied desert, which is why the BLM owns so much land there. Still, I had read online that St. George, Utah was one of the country's fastest-growing cities and had over fifty thousand residents. To my shock, some interns were sent to towns with only a couple hundred people. How bad could my city be by comparison?

My main responsibility for this internship was updating the BLM's water catchment database, along with another intern from the same program. He was a non-traditional intern, approximately in his 50s, having changed his career later in life, but with a lot of outdoors experience, unlike me. The catchments we had to look for were manmade water sources that supplemented wildlife with water during times of drought in the desert. Our job was to spend the summer navigating to over a hundred of these locations in the database and determine if a catchment was really there. When we arrived at the supposed location, we would search the area extensively to make sure there wasn't a catchment, and if we did find one, we confirmed or updated the coordinates.

This work involved a lot of backcountry navigation and I became good at using common mapping software in wildlife biology, ArcGIS®. We had to create our own maps consisting of a series of waypoints that eventually led us to the catchment location. This was way before Google Maps, but even today I doubt that a navigation app on a smartphone would be helpful. We worked in areas with hardly any development and navigated all backcountry dirt roads. I doubt there would be reliable cell service there, even today.

I was disappointed that my catchment research didn't directly involve wildlife, but I did end up getting a lot of field experience and as a result realized how difficult it could be. Many areas where catchments were supposed to be were inaccessible by vehicle resulting in us hiking on foot to look for them. People often say that the desert has dry heat so it's not that bad compared to lower temperatures in a more humid climate. However, once the temperature was over 100, it was just plain hot to me. I distinctly remember hiking through a sandy stretch on one of those dry, hot days, seemingly going nowhere. My feet were hot from my hiking boots as they sank into the sand, and I sweated profusely through my clothes, trying to cool myself with wet bandanas around my wrist.

*Taking a coordinate at the northern rim of the Grand Canyon in Arizona.*

Other times, we had to "negotiate" the land, meaning we had to figure out how to get from point A to B. On the surface it looked straightforward and easy, and when we first started, I didn't understand why we even had to plan out where we were going to go. Obviously, we should just hike straight to the waypoint. But once you started walking a bit, rock crevices and narrow canyons obscured from vision would suddenly appear, requiring us to backtrack and start over if we used that approach. At those moments, I remember hating field work. Was this what wildlife biology was really like? At those times, I doubted whether I had what it took to make it in this field.

Despite mostly searching for catchments, I was also able to gain some experience directly in wildlife biology and conservation. A few times, I accompanied a young wildlife biologist conducting surveys for the Southwestern willow flycatcher, a federally endangered bird species. Being greyish-yellow with black wing tips, these birds don't look particularly special or distinct; rather they look pretty drab. The differences between this subspecies from other flycatchers are so subtle that identification by bird song is best. We slowly trenched through water and dense vegetation taller than us, playing the calls of a male on a recording device along the way. As these birds are territorial, a male flycatcher will come towards the speaker and sing to defend his territory, thinking another male is in the area. This method helped us determine if the species was present or not in that patch of habitat.

In this internship, I also had my first experience using radio telemetry, a common method that wildlife biologists use to study animal movement. First, scientists capture wild animals temporarily, anesthetize them depending on the size of the animal, and then attach a tracking device to the animal via a collar, backpack, adhesion, or even implantation. The tracking device gives off a signal and after they release the animal back into the wild, the scientists collect location coordinates of the animal using an antenna that picks up the unique signal each tracker gives off. Over time, the scientists collect enough location points to understand where the animal goes and to generate an estimate of the animal's home range.

I participated in two different radio telemetry projects. In the first project, we tracked the locations of fledging birds (birds that were starting to leave the nest) and recorded their behaviors. The trackers were not as much for studying the birds' movement patterns but were more for ensuring we could find the birds every day to observe them. This was a project of major interest to me, and definitely more along the lines of what I wanted to do for a career, but we were only helping out for a few days.

The second project centered on understanding the movement patterns of bats in the northern rim of the Grand Canyon. A small team of us headed out to the north rim late in the afternoon one day to set up mist nets for capturing the bats as they would leave their roost that evening. These nets look like draping volleyball nets but are gentle and designed to capture birds and bats. When one flies into the net, they become tangled and need a professional to take them out.

We captured six bats that night. The wildlife biologists affixed the transmitters to the back of the bats using a temporary adhesive and wrapped them in a tiny cloth, a "bat burrito," to make sure it was secured. The wildlife biologists released the bats, and we tracked them throughout the night sleeping in tents close (but not too close) to the canyon. After a few nights of tracking bats, back at the office, I imported the coordinates into ArcGIS® and looked at the locations throughout the canyon, trying to make sense of the flight patterns.

This internship was my first taste of real wildlife biology fieldwork, and to be honest, I almost didn't make it the whole way through. In fact, en route to Mount Trumbull, one of our overnight trips in the field, I silently broke down and let it all get to me. I was incredibly lonely, bored any time I wasn't working, and not passionate about my catchment fieldwork. The second we reached the field station near the top of the mountain, I called my dad knowing it would be one of the few areas where we would have enough of a cell phone signal to make calls. I cried to him telling him that I wanted to come home.

In St. George, Utah, I learned that where I lived played a large role in my overall happiness. It was my first time moving away from home and living on my own. Our BLM office was made up entirely of older employees who had families. After work, they went home to their children, and during the weekends, they organized family outings. I was single and without friends, which made me feel awkward going to places alone. Even if I did have a friend to go with, there were few places to meet people and not a whole lot to do.

I felt almost as much culture shock in Utah as I did when I visited Kenya. The influence of the Mormon religion meant the town was extremely conservative. There was only one bar that everyone advised me was too dangerous to go to, and most restaurants did not even sell alcohol. Not that I needed alcohol to have fun, but back in Buffalo, bars and coffee shops were where I hung out with my friends and how I met people before the Meetup app or social media. In St. George, almost everyone was Mormon, which meant in addition to alcohol, they weren't allowed to drink coffee, and some were even prohibited to watch R-rated movies. Even if I made friends, I was really limited in the types of activities I could do with them. St. George was surrounded by red rock desert, but it was so hot that it was difficult to spend time on the weekends outside. As a young woman, I also didn't feel comfortable hiking on my own, and I was already hiking a lot during the work week for my job, so I craved different forms of entertainment.

I also underestimated how small a town of 55,000 people was for me. I understand this can be large for some people, but I grew up in the suburbs of Buffalo, NY. Although Buffalo is not that large in itself, if you drive 20 minutes in any direction, you hit a new suburb with new stores, restaurants, and movie theaters and therefore things to do. In comparison, St. George wasn't even large enough to have a full mall. There was a JC Penny's and a Target, but that was really it. Going to Target every weekend became my main source of entertainment. If you drove even 10 minutes outside of town, you would find yourself surrounded by desert. As a result, I went back to my apartment after work every day, watched television, ate out of boredom, and eventually became depressed.

Although I badly wanted to quit, I didn't because I worried it would jeopardize my ability to get future jobs. This first internship was difficult enough to get, and I knew if I didn't have a good reference, my next job would be even harder, or potentially impossible to secure. I persevered and stuck it out for the remaining months, which felt like forever, until the internship ended. I was unable to line up another internship or tech position right away, meaning that I had to move back home to Buffalo and continue my search for my next experience there. I was so excited to move back and see my friends. My intuition told me my next job would be better.

I don't remember how, but somehow, I came across Disney's advanced internship program. When I was still in college, I remembered seeing the flyers advertising the regular internship program in the halls of my university and even went to the seminar explaining the program. Having been a huge Disney fan growing up, it was something that interested me, even if it was outside of any of my career goals. Disney World was the premiere vacation destination for our family. While other kids at my school went to the Caribbean or Florida year and year, our family saved for years to splurge for Disney World. It was our favorite place to go. When I went to the informational seminar about the program, I learned it mostly consisted of working a semester in Disney World at an entry-level job such as a ride operator or vendor in the gift shop. This certainly wouldn't help my career as a scientist, and even though working at Disney World sounded exciting, working in those types of positions did not excite me to apply to the program.

The advanced internships were different though. Most of these internships were business-oriented and intended for those who went through the initial internship program. In fact, you had to have been a Disney intern to participate in the advanced internships. However, there were exceptions in some advanced internships if you had a specialized skill. Science was one of those. Disney offered dozens of advanced internships across different scientific disciplines such as animal behavior, captive animal management, biotechnology, environmental education, and crop science. This was quite honestly the perfect internship for me.

I spent weeks working on my application, making sure it showed off my credentials as a young scientist, but also reflected my extreme enthusiasm for the Disney corporation. It worked and I received an interview. I scoured the Internet for examples of interview questions and wrote out answers to every single one on 3 x 5 cards in Sharpie® markers. Most were related to business and not wildlife biology, so I had a difficult time finding appropriate questions and advice but still tried my best to adapt them to my field. When the time came for my phone interview, I splayed them across my desk in the basement so I could refer to them if I froze in the moment. I was nervous, but also super excited.

I interviewed with five people who worked at the *Wildlife Tracking Center*.

This was incredibly intimidating, but they were all friendly and conversational, even joking at times, calming my nerves. When the interviewers asked me the last question, what my favorite Disney movie was, I knew my answer right away. I explained that *The Little Mermaid* made the oceans come alive for me and I had always loved how Disney movies tried to accurately portray species and ecosystems. As the interviewers complimented me on my choice and chatted briefly with me about the movie, I had a good feeling about the position. These were my people.

I was right about my good feeling in that I was offered an advanced internship, however, I didn't get the one I applied for. Still inspired by the elephant researcher in Kenya, I really wanted the coveted animal behavior internship, which I found out later, was incredibly competitive. Instead, I accepted an endocrine and wildlife research position at the *Wildlife Tracking Center* in *Disney's Animal Kingdom*. *Animal Kingdom* is part theme park with rides, stage performances, and parades, but also a fully accredited zoo with over 2,000 animals across 300 different species. My research largely focused on studying the hormones of captive animals at the zoo for husbandry and management. My experience running PCRs and gels in the flour beetle lab paid off; it was proof that I would be great at performing the hormonal assays involved in endocrine research.

Zoos frequently want species to breed as part of the *Association of Zoos and Aquariums' Species Survival Plan* to ensure a genetically diverse captive population should wild populations crash and need species reintroductions. Zoos need to know information like a female's cycle to determine when they should place males and females together and when younger individuals start their reproductive cycles. If animals were pregnant, we monitored their hormones to inform veterinarians when the animals were close to giving birth. At the time, we monitored a pregnant elephant very close to giving birth. Elephants are pregnant for approximately two years and zoos have struggled to breed elephants successfully in captivity. Our job was to run the elephant's blood samples and look for signals in the hormones that indicated the calf's birth within the next few days. This information gave time for the vets to prepare for the birth and increase the chances of successful delivery.

Other species breed well in captivity and are given contraceptives. Another major part of our job was monitoring how effective contraceptives were at suppressing females' cycles. Much of this research focused on the cotton-top tamarin, which ironically is an endangered species in the wild. We also processed fecal samples sent to us from wild cotton-top tamarins in their native range of Colombia to study their reproductive cycles for conservation purposes.

I spent most of my days in the lab, pipetting solutions into plastic rectangular "plates" filled with small holes. Our lab looked quite different from the flour beetle lab; it was decorated with animal paw print stickers leaving tracks all over the benches and walls. Stuffed Micky Mouse dolls with safari hats sat near the computers and next to the lab equipment. We worked behind a glass wall, so guests could see the research side of *Disney's Animal Kingdom* by watching scientists at work. I worked with a great team of supportive scientists and during lab meetings we read team-building books. We even did activities outside of work together like learning how to scuba dive. For our certification test, we spent a weekend scuba diving off the coast of Florida.

*Pipetting in the Wildlife Tracking Center Lab at Disney's Animal Kingdom.*

Additionally, I had a built-in friend group through the larger Disney internship program. We all had to take classes together to learn about park protocols and quickly bonded. We lived in the same apartment complexes reserved for Disney interns and ate lunch together at the staff cafeterias. As employees of Disney received free entry into any of their four theme or three water parks, there was always plenty to do. My experience in the Disney internship was the exact opposite as my internship with the BLM. To this day, *Animal Kingdom* was the best place I have ever worked. From this experience, I not only learned the importance of where you work, but who you work with. The positive work environment and collaborative colleagues made the *Animal Tracking Center* a place I looked forward to going to every day.

In addition to the endocrine research, which I loved, I also worked alongside a wildlife biologist surveying species on undeveloped land in *Disney World*. *Disney World* owns a lot of land surrounding the park that they will never develop, and they hired a wildlife biologist to assess the current status of several taxonomic groups so they could better manage their wildlife in the future. We drove all over on the outskirts of *Disney World* stopping at seemingly random places to conduct point count surveys for birds, where we would record all of the birds we could see and/or hear in a location for a designated amount of time. Some of the point counts occurred in swamps where we waded through hip-deep water in waders amongst cypress trees. We also surveyed alligators by driving around at night pointing spotlights into the canals to look for their red eyeshine just above the water. This reminded me of my nights in Alleghany as a child, except this time it was alligators instead of raccoons, skunks, and opossums.

The internships at *Animal Kingdom* promoted a culture of learning and I was also able to help out with projects on other research teams. I expressed my love and fascination for elephants in my internship application and the director of our lab did not forget this. She arranged for me to accompany another scientist studying elephant vocalizations in *Gatorland*, a theme park/zoo dedicated mostly to the American alligator.

Why was an elephant biologist at *Gatorland*? At the time, he had completed the data collection portion of his elephant research and was beginning a new project on alligators. We went to record alligator vocalizations. Although I did not participate directly in elephant research, during the long car ride to *Gatorland*, we discussed his research at length, and I learned a lot from him about how I should approach a career studying elephants. We also had plenty of time to talk quietly at *Gatorland*, as none of the alligators vocalized over the several hours that we were there.

The director of the lab led sea turtle research at *Disney's Vero Beach Resort* on the eastern coast of Florida. She invited me to go to a conference she was speaking at and also to participate in some sea turtle research. All species of sea turtles are threatened or endangered, and one major cause is beach development, which affects juvenile survival. Sea turtles lay their eggs on land at night and months later the nestlings emerge and make their way to the sea. Lighting from hotels and other establishments on the beach interferes with the young turtles' navigation systems causing them frequently to go in the opposite direction away from the ocean. *Disney's Vero Beach Resort* has special lighting to prevent this from happening.

Scientists survey beaches during nesting months to mark new nests and monitor them throughout the season. When the young turtles emerge, they can help protect the babies from predators and people as they make their way to the ocean. We surveyed the beach looking for female nesting turtles using red lights and night vision goggles so as not to disturb the turtles. That night a giant female leatherback came to the shore and we watched her lay her eggs in the sand, bury them with her flippers, and return back to the ocean like she had never been there at all.

Finally, Disney cares deeply about public education and conservation. Through this internship, I had my first experiences interacting with the public. The interns took classes on how to communicate science from Disney professionals and we implemented this weekly with something we called "magic moments." We stood in front of the research lab with props like shellacked giraffe poop or the collars used on elephants to record their vocalizations. These props sparked guests' interest and we used them to start conversations about Disney's research. In this internship, I discovered I loved talking to people of all ages and from all over the world about animals and scientific research.

I worked at the *Wildlife Tracking Center* for seven months and reluctantly left, headed back to Buffalo again after the internship ended. At this point, I planned on shifting my focus entirely to applying for graduate school programs instead of more work opportunities. My experience at *Disney World* confirmed I wanted to go to graduate school to continue this career.

But I got derailed from my applications when a friend of mine from the Disney internship told me he was applying for the SFS program to be a student. I became jealous, remembering how much I loved living in Kenya and decided on a whim, to apply again for the internship with SFS in Kenya. To be more competitive, I enrolled in an Emergency Medical Technician course at a local community college. Because the Kenya program was located in such a rural area, they prioritized applicants with medical training. The course paid off and I was offered the internship.

In January 2005, I packed my bags to spend an entire year in Kenya. At the time, SFS hired a completely Kenyan staff except for two positions, the Student Affairs Manager and the American intern. These two positions are meant to address the non-academic needs and activities of the American students while traveling abroad. Most of my daily responsibilities were not related to wildlife or even scientific. Instead, I helped run the program by overseeing some student activities, assisting the Student Affairs Manager, and aiding the professors in classroom preparation.

Honestly though, a lot of times, I felt lost and useless. The students were college-aged and didn't need help adjusting to their new lives. I didn't have that much to help out with in class or with the Student Affairs Manager either. When I was a student, there were three American interns and I remembered them driving the students on safaris and leading other nature-related activities. But I was the only American intern on the program now, and there was a new rule in place that American interns were no longer allowed to drive. Even though I had been to Kenya before, I was not as helpful organizing local logistics or leading nature walks compared to the Kenyan interns who obviously lived their entire lives in Kenya and were extremely familiar with the local flora, fauna, and culture.

My first semester was challenging and just like I had in St. George, Utah,

I fantasized about going home. That first semester particularly, students were constantly sick, more so than ever in the history of the program. Our Student Affairs Manager was also new and incorrectly expected me to have medical expertise to treat the students. While I did take an EMT class, we were mostly taught life-saving skills, many of which would only take place in the US such as administering oxygen in as ambulance. But she wanted me to diagnose students, something that I was not qualified and not supposed to do. A couple of times she even doubted my capabilities in front of students.

This, combined with the fact that I was only a few years older than them, caused them to lose trust and respect for me. Soon, they saw me as another student, but I wasn't a student and couldn't socialize with them as if I were one. At night, I would frequently hear them laughing and talking in their bandas, where staff weren't allowed. While they were laughing and bonding, I went back to mine alone. As in Utah, I felt isolated and became depressed except for this time I was surrounded by fifty people on a daily basis. My sister was getting married over the summer and I thought about going to the wedding and never coming back.

I knew in my gut though that I could not quit, at least not without consequences. How could I cut it as a wildlife biologist if I couldn't make it a year in Kenya? What would I say to advisors interviewing me? They would not have faith in me to be able to carry out international research on my own if I couldn't handle it as an intern in an established program. I felt the truth about being lonely and having a difficult boss would surely reflect poorly on me. I also had a once in a lifetime opportunity. Even though I most certainly had challenges, I had a job where I got to go on safari, see incredible landscapes, and be in the midst of elephants, lions, and hyenas at least once every month. I couldn't let the dynamics with my boss ruin this opportunity for me.

Luckily, towards the end of the first semester, things started to change. Almost all of the Kenyan staff had difficulties with the Student Affairs Manager, especially one of the Kenyan interns I worked closely with, causing myself and the other two interns to bond. They saw how unrealistic her expectations were of me and for the intern role in general and we became good friends. I didn't end up going to my sister's wedding because it would have been expensive and quite honestly, I was afraid after seeing my family, I wouldn't have gone back to Kenya. I knew I had to complete my year.

After a few weeks of traveling within Kenya for fun when the spring semester ended, I returned to KBC and started the summer semester with a clean slate and a positive outlook. We had a completely new group of students coming and they wouldn't know about my problems from the first semester. I could start completely over. In the summer semester, something else happened that started to put my year in Kenya on an upwards swing. In the first semester, I bumbled around with no purpose, but at the start of the summer semester, the wildlife professor asked me to help him publish his current studies.

The professor saw that I had a strong interest in wildlife research, and he asked me if I was a good writer and could do statistics. I said yes to the writing question and confidently stated that I could learn whatever statistics that needed to be done for data analysis. He then invited me to help him publish two studies by overseeing the student research projects, analyzing data, and writing the first drafts of manuscripts. At the end of each semester, students conducted final research projects. With four different groups of students collecting data each year, the professor gathered enough data for two publications.

I had a new purpose and having authorship on peer-reviewed journals is one of the best and most important things you can do for a career in wildlife biology. The work involved in the research was perfect for me, and the questions really drew me in. The study actually consisted of social science research and did not involve studies on the animals themselves. It was on understanding human behavior and attitudes in relation to wildlife.

*Elephants in fronts of Mount Kilimanjaro in Amboseli National Park, Kenya. This was a common scene. Photo by Kaira Wagoner.*

We were testing to see how important the "big five" were to tourists visiting Kenya. These are historically the five most dangerous animals to hunt: elephant, buffalo, leopard, lion, and rhino. Many experts in the tourism industry thought tourists chose what parks they visited in Africa based on their ability to see the big five. Parks without some of the big five, like Amboseli, which lacked rhinos, would be at a disadvantage if this were true. But we suspected that the big five was a thing of the past and that today's tourists would be more interested in overall biodiversity, cultural attractions like the Maasai's manyattas, and natural features like Mount Kilimanjaro.

To test this, we had fieldwork where we followed tourist vehicles around Amboseli National park and a survey component to the research where we interviewed tourists at the lodges. This meant we were either on safari for most of the day, following tourist vehicles, stopping and recording which species they stopped for and how many animals there were, or approaching guests in the fancy lodges, asking them the survey questions and writing down their answers. Of course, I treasured this research experience. Who wouldn't?

We spent all day surrounded by picturesque views of zebras, giraffe, and buffalo in front of the lavender-colored hues of Mount Kilimanjaro. When I wasn't on safari, I was having lighthearted conversations with tourists from England, Germany, and France while sipping cold bottles of soda, asking them about which animals they wanted to see most.

Although this "fieldwork" was the best part of my job, the real work was back at the office at KBC. There I did the data analysis and writing for the two manuscripts. Data analysis consisted of a series of chi-squared tests and I had to learn a new computer program called SPSS®. This was my first time analyzing data; in my previous internships, I had only collected data or processed samples and then handed off the data to a more senior scientist for analysis. This was also the first time that data analysis, specifically being proficient at statistics, was emphasized to me to be successful in this career. Dr. Okello told me that if I could master statistics, I would be successful in this career. At the time, he gave me a flow chart cheat sheet as to what type of statistical tests to use under different circumstances. However, for his research, I didn't have to think about what kind of data analysis to do. He already chose the analysis given the experimental design of the study and showed me to run it in SPSS® resulting in little struggle for me to get the results.

I thought the data analysis would be hard, but instead the writing ended up being much harder and requiring more focus. Just to write an introduction, I had to summarize previously collected published results from dozens of papers, which meant a lot of reading of dense material from the scientific literature. Science writing is very different from the writing I was taught in English class. In school, I was rewarded for writing creatively with metaphors and advanced vocabulary, but science writing is all about being concise and to the point.

At the time, I thought my writing was near perfect, and by the time I had left Kenya, I thought I left Dr. Okello with a publishable draft. To my surprise, the papers weren't published until years later (which I later found out was not uncommon), and when I read them, I barely recognized the writing. I was definitely not as good as I thought, and he ended up revising the drafts a lot. At first, I felt embarrassed and questioned if I really was that helpful. Having gone through the process myself now over twenty times, I've realized that science writing is a cultivated skill, and even if you have to revise the writing a lot, having someone organize it in an early draft is indeed helpful. Even when I was in my second postdoc and opened up early drafts from my dissertation chapters to work on for publication, I cringed at how bad the writing had been even with years of experience by that point. The publications from our research in Kenya helped me greatly in my career and one of them is still my most cited publication to this day.

Despite having this glamorous fieldwork without the typical long days and austere conditions under the wildlife management professor, ironically this internship in Kenya gave me a lot of credibility as a tough wildlife biologist. Simply living in rural Kenya for a year was proof in of itself that I had what it took to live long-term in field conditions. In reality, the luxurious fieldwork under the wildlife management professor only made up several weeks out of the entire year after every semester ended.

Most of my days were spent at the field station while students were in classes or traveling for various field trips as part of the program. We would spend weeks visiting different parts of Kenya from Tsavo National to the Maasai Mara in the west and Nairobi in between. While this travel may sound splendid, it involved long days of driving on incredibly bumpy roads, flat tires, getting stuck in the mud, and accumulating large amounts of dirt on your clothes simply by sitting in the car with the windows open from all of the dust. You couldn't relax or sleep because the roads were so bad that you frequently had to brace yourself from being tossed around the car. I also traveled by myself within Kenya and to Uganda and Rwanda in between student sessions almost always using public transportation. My patience grew dramatically during this time as sometimes buses were hours late and almost always broke down at some point along the journey. I learned to keep a book by my side at all times.

Other fieldwork we conducted at SFS with the ecology and human dimensions professors was challenging even though we weren't necessarily doing anything that physically difficult. For example, the students took measurements of the landscape and vegetation, and interviewed the local community using in-depth surveys. These interviews were tedious and required patience by the students and were nothing like those we had conducted by the pool with tourists at the lodges. We interviewed people at their *bomas* (traditional Maasai homes), frequently with little shade, and because most people owned cattle, there were houseflies constantly sticking to your skin (I swear the houseflies in Kenya move slower than the ones in the US). Each question was translated from English to Maa, the language of the local Maasai community, and the answers were translated back. It was not uncommon for a single survey to take an hour.

The SFS program was highly luxurious when compared to most

traditional fieldwork, yet still some students complained and had a difficult time. For example, a full-time chef cooked most of our meals even when we went camping overnight in national parks. The food was delicious and even included some staples from back home like guacamole and pizza. But it wasn't like the pizza in the US and students would still complain about missing US pizza and other food like burgers, French fries, and ice cream. Yet when the students weren't there or I was traveling by myself, I realized how much luxury we had.

Once the students left, the chef still made meals for the staff, but we had traditional and less expensive Kenyan meals like *ugali*, a cooked cornmeal resembling a thick cake of Cream of Wheat®, served with beans and kale or cabbage. We ate this nearly every day. Don't get me wrong, I honestly love traditional Kenyan food and the chef's cooking, but if students were complaining about what they had before, they would have been miserable with the redundancy of the food we had now. Many simply hated *ugali* too. In other words, some people still had a difficult time even with these luxuries and therefore would have really struggled during a regular field season. Once I started to correspond with potential advisors for graduate school and expressed interest in working internationally, they had confidence in my ability to conduct research abroad having lived in a remote area within a developing country for a year.

In Kenya, I also got a taste of what graduate school would be like when I accompanied a professor for fieldwork in Tsavo East National Park. In Kenya, I continued my research on potential graduate advisors from the year before and started to contact those I was interested in working with. My graduate school search was similar to my college search in that I didn't really know what I was doing or what kind of research I ultimately wanted to pursue. I knew that graduate school was mostly about my research, and that the professor I chose as my advisor would be my most important decision, but I had an incredibly vague sense on what I wanted to do. Most people imagine that applying to graduate school is similar to the application process for undergraduate degrees; you find a university and program that you are interested in attending, and you apply through a standard application. That is still a component of graduate school, but far less depends on the formal application. In fact, that part is really just a formality. Unless you are applying for an advertised position, you first have to develop a relationship with the potential advisor. This means you contacted the professor before you apply, and they have agreed to take you on as a student if you are accepted.

My approach was to research almost every wildlife, zoology, ecology, or biology department in the United States (I kid you not). When I began this search in Buffalo, I bought a gigantic book of universities in the US with short descriptive summaries and department information. We did have the Internet then, but websites for this topic weren't nearly as organized and useful as they are today. I opened my book and looked at each university. If the university had a department I was interested in, I went online, looked up the department, and visited the professors' individual research page to see if their research was appealing to me. Remember, I had no idea what I wanted to do. As a result, I had collected the information from professors conducting studies all over the world, in different ecosystems and across diverse taxa. While in Kenya, I contacted one professor who happened to conduct a citizen science program there, and he invited me to join him for a few days to learn about his research and talk graduate school. I took several days off from my internship and headed to Tsavo East National Park to meet him.

His research focused on the famous maneless lions of Tsavo. It was on trying to understand physiologically how the lions were maneless and what were the evolutionary forces driving manelessness in this population. The citizen science research consisted of driving around the park with a radio telemetry antenna and looking for lions that had previously been collared with trackers emitting signals. Because lions are typically more active at night, we started our surveys late in the afternoon and worked until sunrise. This was so exciting for me; I had never done a night safari before. Along the way, we recorded what other birds and mammals we saw and how many we spotted. Because it was citizen science, we had cars full of volunteers from all over Europe and the US all helping out.

We started our first safari at around 5 pm when it was still bright and sunny outside. With antennae in hand, we turned to specific radio signals to try to pick up the beeps of the individual lions. I started the safari standing on the seat of the vehicle scanning the dusty savanna through the open top alongside the professor. I enthusiastically searched and imagined carnivores that I had never seen before becoming active at night and making themselves visible to me. But we drove around for hours without seeing much but zebras and some different raptors like hawks and owls. It seemed like we should be seeing lions everywhere, but we hadn't even seen one for the hours that we spent searching. But we kept our spotlight on and just kept on going.

As we weren't seeing much on our research expedition, naturally, our conversation started to shift towards graduate school and my own personal research interests. He asked me what particularly I was interested in researching, but I didn't know. I gave general answers about how I wanted to conduct research with a conservation application. Even though his study was on lions, mine didn't have to be, and really, I could use that field site to study whatever I wanted. But not only did I not know what I wanted to study, but I also had no clue what was interesting to research. Scientists had been conducting research consistently in Kenya for decades, surely, I thought, most of the animals had been well-studied.

Quite honestly, I expected him to help me figure out my research questions by presenting several projects for me to choose from. No one had asked me to come up with a research question before. I didn't know what was worth asking. Couldn't I just watch animals and record their behaviors? Make observations about what they were doing? Wasn't that what Jane Goodall did? Or wasn't there some sort of authority that listed out interesting and important things to study for the conservation of that species? My mind was blank. My courses in school always centered around a series of concepts, facts to memorize, or vocabulary to learn. I wasn't taught how to ask scientific questions. In fact, teachers would sometimes get annoyed when you asked too many questions.

I felt defeated and stupid. Maybe I wasn't ready for graduate school. That whole night as we drove around searching for lions by spotlight, my mind raced trying to think of what research questions I could ask, but I came up empty-handed. As the night went on, I started to get tired from standing nonstop and looked for lions from the side window while seated instead. "Don't fall asleep, don't fall asleep," I repeated in my head over and over again, but it was inevitable. Despite all of the coffee I drank to keep myself up, I couldn't help it. My fatigue got the best of me and I found myself drifting off, waking up from large bumps in the road, and then falling right back asleep. In college, I was never successful at pulling an "all-nighter." I knew I did not make a good impression.

The next day I woke up trying to avoid feeling defeated from my interactions with the professor the night before. Since we were up all night, we were free to relax throughout the day and catch up on sleep. Again, I pondered over what my study questions could be, so I could impress him later on that day, but didn't I come up with much. I decided to be interested in the maneless question for the time being because I still struggled to come up with any questions besides the ones he presented. However, I wasn't super excited about this question. Although I thought it was interesting, understanding why Tsavo lions were maneless didn't have any conservation implications or application; it would be a purely scientific study and I was driven by conservation.

That night, we repeated the same fieldwork, leaving in the late afternoon and staying up until morning looking for lions with our antennas and surveying wildlife along the way. This whole time we had been scouring the savannas, intensely searching through every patch of dry, thorny vegetation to see if we could make out anything lion-like. Later that afternoon, we turned the corner to a large open area near a watering hole and saw one of the famous Tsavo lions laying out in the open on a small mound of soil. Her head was up and facing our direction, almost as if she was waiting to be found by us. We stopped the vehicle in front of her and she watched us park, moving nothing but her head.

The lioness had no collar. Therefore, she was not one of the lions we were tracking in the study. She was lightly colored, almost white, like beach sand,

and stuck out starkly against the dusty, dirt-covered backdrop. Lions don't need to hide unless they are hunting; they are the top predator in this ecosystem. I had seen lions before, but seeing lions is always special and a jaw-dropping experience. We stayed with her until sunset and recorded data, watching which direction she would move in. But after over an hour of waiting, she never left her spot, and we had to move on.

Night two was similar to night one. I struggled to come up with interesting conversations about research and preferred to just absorb knowledge from the professor and learn. I think he sensed I was intimidated and unsure of my next steps and stopped probing me about my potential research questions. This was my first real time feeling impostor syndrome in my wildlife biology career. Impostor syndrome describes the feeling of not belonging or feeling good enough despite your qualifications. While I certainly didn't know everything when I started any of my jobs, I never felt like I didn't belong, or I wasn't qualified to be there. This was the first time I thought that maybe I wasn't good enough for this career.

I pushed those feelings down inside me though and continued contacting professors for graduate schools the following year. Luckily, I found a few supportive professors that encouraged me to apply for positions in their lab and a few others not as enthusiastic, but also not saying no. I felt especially happy receiving an encouraging email from a professor who was just starting in her new position and conducted forest elephant research for her Ph.D. The professors I was interested in were from all over the country and there was not a lot in common amongst their research programs. I was interested in research ranging from sea turtle migration patterns to zebra behavior. The common feature was that these programs tended to be on large vertebrates with a conservation application.

I ended up finishing out the rest of 2005 in Kenya on an upward trajectory and positive note. I could write a full book on my experiences and lessons learned in Kenya alone. Overall, like my internship in St. George, I was grateful I stuck it out for the year during those difficult times. My overall experience got much better and I now could add two highly coveted skills to my resume: co-authorship on a manuscript and working in a developing country for a year. After I left Kenya, I started to feel like a real wildlife biologist.

# 3 LEVELING UP WITH A PH.D.

Leaving hot and sunny Kenya behind, I went back home to a frigid Buffalo in December and lived with my parents again. During this time, I finished up my graduate school applications and submitted them. Then I waited several months for the results. Because I wasn't sure of my chances and what I really wanted to do, I ended up submitting applications to ten programs including some where I did not have that much or any support from the professor. Albeit expensive (each application cost at least $100), the more graduate schools I applied to, the higher my chances were of getting in.

In spring 2006, I received mail indicating that two universities accepted me into their research programs: The University of Missouri in Columbia, Missouri and Texas Tech University in Lubbock, Texas. I ended up only getting into the programs where I had a lot of previous correspondence with the professor.

The professor at Missouri allowed me to pursue whatever research interest I wanted to, as long as it contained a genetic component. Her research focused on the application of ecological genetics from non-invasive samples (e.g., dung, scat, hair) to answer questions about species' gene flow, population structure, and the relationships between individuals in the population. Additionally, the Biological Sciences department there offered me a four-year fellowship in their Ph.D. program. This fellowship included a higher stipend and I would not have to be a teaching assistant (TA) during those years, which would allow me more time to focus on my research.

At Texas Tech, my research would have most likely focused on fish, which was not my original intention. I applied to work with this professor because I was intrigued by the dolphin research she listed on her website. When I visited Texas Tech though, she tried to persuade me to study fish because the dolphin research was more of a side project and not as active as her other research. Despite fish having more direct conservation implications because people around the world consume them, I was not as excited about studying fish (sorry fish). This professor also encouraged me to get a master's first before a Ph.D. I would therefore have to be a TA every semester, meaning I would have a lot less time to focus on my research due to teaching responsibilities, and more work overall. My end goal was never to become a professor, so teaching a laboratory as a TA was not very alluring to me or helpful to get me to where I wanted to be in the future of my career.

Before I made my decision on where to go and what degree to get, I visited both schools, but it was really difficult to compare them. The Biological Sciences department at the University of Missouri invited potential students to visit during a fully scheduled, official recruitment weekend while at Texas Tech, I visited on my own. In the formal recruitment weekend at the University of Missouri, the university arranged for me to fly in to visit campus and spend two full days in interviews with the department's professors and students, exploring the campus and Columbia, and attending faculty and student parties. Both days were jammed packed with activities and interviews. During the recruitment weekend in Missouri, I interviewed with almost all of the professors and many of the students from their labs in the Behavior, Ecology, and Evolution subdepartment within Biological Sciences.

Because choosing the right advisor and graduate program was important to me, I paid out of pocket to visit Texas Tech University unofficially. Even though this was a financial burden, I believed it would be worth the cost to be confident about my decision. In my unofficial visit, I spent a lot of time with my potential advisor, toured her research lab, and went out to lunch with her students. I felt like I had a solid understanding of what it would be like to work with her, the lab dynamics, and what I could conduct my research on. I was, however, not able to interact with other professors or students from other labs as I had been able to in my formal recruitment weekend at the University of Missouri. Not having met more faculty and students made it difficult to compare schools, but by and large, your advisor is the most important choice in your graduate program, and I weighed my decision mostly based on their research. In the end, I chose the University of Missouri because of the benefits from the fellowship and I was more excited about the research I could do there.

In August 2006, I moved to Columbia, Missouri, and started my Ph.D. in Evolution, Ecology and Behavior within the Biological Sciences department at the University of Missouri. I enrolled in a few courses, but really most of my mental energy was focused on what I was going to study in my Ph.D. program, my highest priority and the thing that would affect the rest of my life the most. When people think of school, they automatically think of taking lots of classes and studying for exams, but in graduate school, there is far less emphasis on your courses than your research. The courses are also structured differently with more emphasis on discussions and projects incorporating your research, and less on studying and tests. In fact, I didn't take a single exam for any of my classes in graduate school. Graduate school is much more like having a job. Day in and day out, your main responsibility is to conduct research for the purposes of completing your thesis or dissertation.

I spent the first six months of graduate school trying to figure out exactly what I was going to study. To develop my research objectives, I read a lot of scientific papers, especially on African forest elephants (*Loxodonta cyclotis*). My advisor was an expert on non-invasive genetics using dung and published many papers on this species. Despite their large size, forest elephants are quite difficult to study because they are almost always obscured from sight by the rainforest vegetation that they live in. Unlike the savanna elephants (*Loxodonta africana*) that you could see from miles away in the distance that Cynthia Moss studied, you often can't see forest elephants even beyond a single tree or bush in front of them if they remain still. Using DNA from their dung is one of the best and few ways to study them.

I definitely bumbled around for the first six months, reading papers about using genetic analyses on animals ranging from rhinos to kinkajous, waiting for something to spark the perfect research questions for me. I felt lost and honestly unworthy without a research question and started to get feelings of impostor syndrome again. When I took out seminar speakers to lunch, and they asked me about my research, flashbacks of being asked what I wanted to study by the professor in Tsavo East came back to me and like then, I had no response.

I always did so well in school; as long as I studied hard, I got good grades. But grad school was already different. The process was nebulous and vague. I read paper after paper, afterward thinking that it was an interesting study and something that I could do. Why couldn't I think of something like that? But I didn't have any original ideas for myself. I was struggling to figure out what was something that had never been studied before, was interesting, but could be accomplished within several years. Being more than several months into my Ph.D. without research questions, I started to panic.

Finally, I got the spark I needed. My advisor put me in contact with a Wildlife Conservation Society scientist working on forest elephants in Gabon. He was getting some unexpected results from a population of forest elephants in Loango National Park. Six elephants were fitted with collars with GPS tracking devices in the park. When he and other scientists began the study, they thought the elephants would migrate large distances between two national parks in the area, but the GPS data from these elephants showed, surprisingly, that they never left the park they were collared in. The elephants had incredibly small home ranges, some being only several kilometers, and they didn't overlap much with each other either. Based on the raw telemetry data, these elephants were clearly doing something different than most other

African elephants.

The female African savanna elephants that Cynthia Moss studied live in complex fission-fusion social systems where core family groups typically stay together, but they meet up with extended family over space and time. The group sizes change according to the wet and dry seasons, but eventually, almost every female ends up interacting with every other one in the population. Some elephants have huge home ranges, for example, an upwards of tens of thousands of kilometers in the deserts of Namibia, and studies with multiple elephants from the same population showed overlap between individuals' home ranges. Less research had been conducted on Asian elephants (*Elephas maximus*). Their group sizes are generally smaller, but they could still be sizeable, with larger, overlapping home ranges.

The results of this scientist's forest elephant movement study inspired me to investigate the social structure of this species and compare it to the other extant elephants, African savanna and Asian, to understand their sociality in light of their ecology. Elephants initiated my interest in pursuing wildlife biology as a career, so investigating the social structure of forest elephants seemed like a perfect choice for me. I used three different approaches to better understand forest elephant social structure.

The first chapter (i.e. section, explained later on) of my dissertation was to formally analyze the GPS location data from the Loango elephants. I was going to estimate home range size and how much the elephants overlapped with each other. Somewhere in Libreville Gabon, there were blood samples of the elephants on filter paper. Because savanna elephant family groups are composed of related individuals and share the same matriline, it was my mission to get these forest elephant blood samples, extract the DNA from them, and see if the individuals with the most home range overlap were also the most related.

My second chapter involved research methods that were very similar to Cynthia Moss' methods. They consisted of observing forest elephants in the field, identifying individuals, and recording which individuals were seen with each other, just like how the researcher in Kenya described to us. Forest elephants are difficult to observe in most of their range, which is why they are so understudied, but there are some parks that have patches of savanna habitat or "*bais*," open areas rich with salts and minerals that attract animals. As long as I carefully chose my field site so that I would be able to see the elephants, I would use the collected observational data to create network models of the elephants' interactions over time, providing a clearer picture of their relationships.

Previous observational studies in such areas suggested that forest elephants are not very social; the most common group observed was an adult female and her calves. A mother with her calves is not really a group at all because the calves are completely dependent on their mothers until they are older. However, I thought that forest vegetation surrounding the open areas could still be obscuring other elephants from view and that there was more to the story.

After viewing the elephants in the field, I would go on foot to the location where the elephants were to look for fresh dung, the source material for genetic analyses. Using microsatellites and mitochondrial DNA, I would use these dung samples to estimate how elephants were related to each other and identify their matrilines. In the field, I would also measure the size of the elephants' dung bolus to generate a rough approximation of age (adult or juvenile).

For my third chapter, I would examine relatedness patterns of forest elephants across the landscape without needing to see the elephants. With a team of field assistants, we would cover different sections of the park, again searching for fresh dung samples. If forest elephants lived in family groups like savanna elephants, I predicted dung collected closer to one another and of similar freshness would come from more closely related individuals.

In the summer of 2007, I visited potential field sites in the Republic of Congo and Gabon to conduct my research. I needed a park where I could see forest elephants clearly for individual identifications and there are only a few parks that have either *bais* or fragments of savanna habitat. My first choice was Ivindo National Park, Gabon, a forested park containing Langoué Bai, the largest *bai* in Gabon. It was perfect. In the middle were small pools of water and soil surrounded by several hundred meters of short grass in all directions so you could see animals clearly. Elephants visited the bai nearly every day and there was already a large viewing platform built for ongoing gorilla research. The *bai* was a few minutes' walk from an ecolodge that I could stay at while I was conducting my research.

*Views of forest elephants from the platform at Langoué Bai.*

I was so excited about working there, but a few months after getting permission to work in the park, park officials reversed their decision. Elephants visited Langoué Bai every day and because I needed to collect dung, they were concerned that my physical presence in the *bai* might disturb the elephants and potentially prevent them from returning. The elephants were their first priority. Quite honestly, I was a little relieved because I started to become nervous about going into the *bai* to collect dung. I would obviously go when there were no elephants present, but nothing would stop elephants from entering the bai at any time while I was still in it.

At the time, I felt like my research was falling apart before it even began, but luckily, I had a backup location that ended up working out even better. During the same scouting trip from before, we visited Lopé National Park, a park with many smaller *bais* scattered throughout, but also large patches of savanna in the northern part of the park.

In the fall of 2008, I traveled to Lopé National Park, Gabon to begin four months of my dissertation fieldwork. I stayed at a field station in the midst of the park with the field station director and field assistant. Other researchers would come and go, but for the most part there were only a few people there at any given time. Like the field station that I stayed at during SFS, the Lopé one was simple, but had more amenities than most other field locations like electricity during the day, refrigeration, and an office to work in that even had air-conditioning to protect the computers and documents from humidity. With all of my experience living in Kenya, I felt comfortable and capable of conducting fieldwork in Gabon and ready to begin.

Given what I had observed in Langoué Bai, daily visits of forest elephants visiting the rich mineral-filled pools of water, I thought it would be best to start my observations at a *bai* in Lopé to create a similar research setup. For the first week, I went with a field assistant to scout out known *bais* across the park. These *bais* were almost unrecognizable as *bais* when compared to Langoué Bai. Langoué Bai is an impressive, gigantic gap in the forest several hundred meters across, while the *bais* in Lopé were small, looking more like dried up ponds and only ranging in the dozens of meters across even for the biggest ones.

I did find one that I thought could work though. It wasn't really a bai, but a mineral lick of some sort. In the forest, it was an open stretch of a stream that seemed to attract a lot of elephants. Looking at the banks of the stream, you could see tusk marks chipped into the sides where elephants dug up soils to ingest for minerals. The stream had a beach-like bank too and there was old dung scattered throughout its shore. In fact, the day we visited it, a mother elephant and her calf were in the stream.

Langoué's platform was built as a stand-alone platform with several levels and large open areas to work and even store equipment. My platform location depended on finding the perfect tree as it was too expensive to create a stand-alone platform. Therefore, we had to find a tree with a crevice that we could build the platform into. The only tree that worked within the vicinity of the stream bank was a thick, straight *Detarium macrocarpum* tree, which also provided fruits that elephants love (although they would not be ripe at the time of the study).

The branching where the platform would be nested was roughly 60 feet or 20 meters high. We hired a local carpenter to build the platform and a team of field assistants carried in the lumber down the path into the forest. The carpenter was around sixty years old, but scaled up the tree with ease, drilling boards into the tree trunk one by one, making his way up to the crevice with impressive speed. He built a platform roughly six feet across and wide with a plastic covering over the top to provide some shelter for me from the rain. I was so excited to start my fieldwork and watch elephants each day.

The platform looked perfect; it was high above the elephants in the tree canopy. But once I started climbing it, I realized it was a lot scarier than it had looked from the ground. The wooden planks of the ladder drilled into the tree were completely vertical at some points requiring upper body strength to pull me up. As I climbed up and descended from those areas my arms and legs shook from both fear and from engaging my muscles. Halfway up at about thirty feet, I was high off the ground. I looked down once while climbing and made sure never to do it again.

When I finally made it to the top of the platform, I didn't feel that much better. It was incredibly high. And while I was safe as long as I sat on the platform, I kept on imagining different scenarios where I could fall off. Sometimes I would catch myself wondering what the hell am I doing? I am going to be left alone in the forest for seven hours sitting on a small platform high in the trees. I did have a radio to communicate to the field station with, but still, things could happen. I suppressed my fears by thinking about all of the incredible animals I would see from my platform. In addition to elephants, there were gorillas, chimpanzees, leopards, and mandrills in this forest.

Every day I climbed up into my tree with binoculars, enough food and water, rain gear, my camera, and field notebook. I sat and just waited. I waited for hours and hours. Every time I heard a large crashing through the forest, I was certain it was elephants, but it always ended up being monkeys. It was surprising how much noise they could make by simply jumping from branch to branch.

The true activity of animals in the forest did not match up to what I had imagined before I spent my time on the platform. While there were monkeys every few hours and short stocky-looking antelope called duikers passing by, I never saw any other mammals for days and days. To pass the time, I spotted and identified birds, improving my life list. But even that could be difficult as it was hard to see them through the horizontal layers of tree canopy.

Finally, one day, I found what I was looking for. After days of sitting on the platform and hours that day, I heard some branches move slightly beyond the stream. In an instance, a forest elephant and her calf emerged from the vegetation. Much quieter than the monkeys, I didn't hear them coming and then suddenly they were in full view. I wondered how long they had been there without me noticing.

*View from my platform of the forest elephants that visited the stream.*

I watched them quietly from above and they gave no indication that they knew I was there. The elephants in this park have a characteristic behavior when they suspect a human is nearby. They extend their trunks upwards above their own heads to gain height and smell farther than their immediate surroundings to confirm it is indeed a human. But these elephants seemed calm and unaware that a human observer was watching their every move far above them. They stayed at the stream bank for hours, digging up the soil, drinking water and pulling fruits off of the trees to eat.

For a few minutes I panicked about what I would do if they didn't leave the area. What if when my field assistant came to get me, they refused to leave? It's not like there was any way that the assistants could get them to leave. They didn't carry guns and shooting guns into the air were not a guarantee the elephants would flee anyway. During my first camping experience in Kenya, lions visited our campsite because they were attracted to some water that was flooding around the outdoor bathrooms. The Kenya Wildlife Service guards that were with us shot their guns in the air. I expected the lions to flee immediately, but they just lingered. They didn't even seem to flinch. It took several shots to finally get them to slowly saunter off. My mind went back to thinking about the current situation. Would this mean I would have to sleep overnight on the platform?

Luckily, the elephants eventually sauntered off too. Ecstatic from my observation, I continued on the platform for a total of about two weeks, but my enthusiasm faded when I realized that was my only observation during the entire time. I did the math quickly in my head. One elephant group every two weeks for the duration of my field season (four months) would definitely not be enough data for my dissertation. This stream bank was not attracting the number of elephants that I needed, and I was forced to quickly change plans.

We decided to switch strategies and focus on finding elephants in the savannas. We were already having some luck with elephant observations there when we had to drive through the park to get to the platform and pick up and drop off the field staff in town where they lived. Along the way, we would sometimes see elephant groups, in which case I collected opportunistic data. While I felt guilty abandoning the platform after all of the hard work from building it, I had no other option. A part of me was also secretly happy that I never had to climb that high ladder again.

My days following the platform largely consisted of driving around the park looking for elephants in the savanna fragments before sunrise and sunset. I loved it. When I saw elephants, I stopped and took as many photos as I could of each individual elephant. Scientists identify African savanna elephants by their ear tears, but forest elephants frequently have very small ear tears or even none at all. Back in Missouri, I pictured myself being able to identify elephants in the field as I saw them like how that Kenyan researcher knew the elephant in the far distance by name at *Ol Tukai Lodge*. I thought once I started to know a few individuals, that I could drive up to a group and recognize some of the elephants instantly.

But when I saw elephants, they would frequently run from me, only giving me a glimpse of one ear or a backside. Despite being in the "open" savannas, the grass could be quite tall, sometimes up to the tops of their heads, still obscuring large portions of the elephant. Even the more tolerant elephants that I could see well could be challenging if they only showed one side. Rapidly, hundreds of photos of different parts of elephants piled up from my observations, quickly overwhelming me. I abandoned the notion that I would get to know this population during my field season.

Days, weeks, and even months later, I sorted through the thousands of photos to identify individuals using a combination of ear tears, ear vein patterns, tusk length and shape, and tail brush pattern. Without seeing complete elephants every time, it became tricky to match up the bits and pieces of elephants I had photos of. If I saw one elephant's left side and another elephant's right side who was of same size as the first elephant, were these photos of two different, but similarly sized elephants? Or different sides of the same individual? Sometimes it was even hard to keep track of individuals within the same group as they weaved in and out of the tall grass and camera's range trying to get ripe fruits to eat. Like Cynthia Moss, I created note cards of the different elephants and named new individuals I saw.

After I observed the elephants in the field, I returned to the same area with a field assistant to search for fresh dung samples. Elephants defecate a lot, up to 20 times a day, so I thought for sure collecting enough fresh dung samples as a source of DNA would be easy. For our safety, we had to make sure the elephants left the area as the elephants in Lopé are notorious for being aggressive.

I envisioned observing elephants in the field for thirty minutes to an hour, watching them leave, and then collecting the fresh dung samples the elephants defecated. However, once the elephants found an area they liked, unless they were just walking through, they typically stayed there for hours. Because I only had a field assistant during the day, if the elephants stayed later than when the workday ended, we couldn't search the area until the following day. The drive to Lopé town to drop off the field assistants was not far, but took some time to drive to, so the workday was shortened.

After my field assistant left, I drove around the park by myself at dusk to look for more elephants. Elephants can be active at any time of the day, but in this park, they seemed to be most active at dawn and dusk. For the elephant groups I saw at dusk, I obviously had to search the area for dung the following day.

My field assistant and I visited all of the places the elephants were the previous day. The areas were all within a walkable distance from the road, but the elephants made navigating the landscape look easy. As we descended from the truck, I left with optimism, but it quickly faded for most of the areas we visited. As we got closer to where the elephants were, we usually had to navigate through grass as tall as us and in swamps. Some of the areas were impossible to access due to the deep trenches in the swamps that remained unseen from the car. I had to constantly watch where I placed my feet; one wrong step and my foot sank straight down, sucking my muck boot deep into the mud. The savanna areas looked like an unkempt lawn from the car window, but once we were close to where the elephants were spotted, I felt like my body shrunk as everything else now appeared gigantic: the undulating landscape, the grasses, and the depths of the swamps.

Sometimes it was incredibly hot and one time I even got heat stroke. Even though we only walked about a quarter of a mile from the car, when we searched the savanna area extensively for dung in late morning or early afternoon and collected samples, this took some time. Sometimes it took over an hour, especially if we found samples, we had to collect them and record data. I got overheated and felt nauseous when we went back to the car and had to lay down.

My expectations were that once I hiked to the area the elephants were spotted in, I would see shiny, fresh dung in the trampled grass. But again, once we were in the area, the site was much bigger than it seems. Old and freshly stomped elephant trails were everywhere. I kept following ones that were obviously old to my field assistant. In my head, I sometimes second guessed his judgement, but I quickly learned not to after seeing that the trails he followed more frequently resulted in fresh dung while mine were always empty. Given how much elephants poop, I thought the areas where I had watched the elephants for hours would be loaded with dung, but we scoured the area and many times found nothing. Shockingly, about two-thirds of the time, we came up empty-handed. I later speculated that elephants seemed to poop together. I have no idea if this is true or not, but it seemed like if we found one dung sample, we were more likely to find others.

Even when we did find dung, it could be confusing if it was from those elephants. Again, my expectations were different than reality. Dung degrades quickly over time and it is pretty easy to tell very fresh dung from dung that is days or weeks old. Fresh dung has a sheen coated over it and a fresh scent. But once I saw the dung samples, sometimes it wasn't so obvious. If the sample was in the shade, the sheen could last longer, but if they were in the sun, they could dry out and look old. Samples that were fresh could look older when rained on and vice versa.

The park-wide survey portion of my research though proved to be the most challenging, both physically and logistically. I started my planning before I left for my field season back in the lab in Missouri. The director of the field station gave me a map of the national park with 1 kilometer by 1 kilometer gridlines. I used these squares to plan out how many days and teams of people it would take to cover the park in a week. Time was of the essence for this portion of my research. We had to collect dung samples as fast as we could so that the elephants wouldn't have moved around that much to capture spatial genetic patterns without seeing individuals. Because much of Lopé is forested, this information would help us determine if the grouping patterns we were seeing in the savannas were similar to those in the forest. Only a small percentage of Lopé was savanna habitat.

When I sat down with the field station director to go over my plan, I quickly found out that what I proposed was impossible without much more manpower and equipment for which I did not have the budget for. What I thought would be a day-long hike to the west of the field station was really a two-week camping trip. Only the northeastern part of the park with savannas was accessible by car; the rest of the park required us to hike on foot. No matter how adept one was at hiking, it would be slow because there is no system of marked trails to use; it's all wild. We would have to find elephant trails and bushwhack our way through dense parts of the forest with machetes. As a result, I had to greatly scale back the amount of land we could cover in our survey. Although I was disappointed that I couldn't capture the spatial extent that I wanted to, I was still hoping that I had enough of a landscape to work with to find patterns.

*Collecting small pieces of elephant dung for DNA analyses.*

Soon after I recovered from this setback, new ones presented themselves daily. After weeks of organizing for the surveys, I woke up on the first day anxious, making sure I had an effective plan for each team. The first day of fieldwork, though, to my great disappointment, it rained all day, clearing up only after the staff had departed. It was against the field station's policy for field assistants to work in the rain because it's too dangerous in the forest; there could be falling branches and you can't hear as well during rain making it difficult to tell if elephants or buffalo are nearby. Luckily, we still could drive around the savannas to look for elephants. I used this as an opportunity to continue my other work and we actually found quite a few fresh dung samples in the road that we could use in the survey data.

Throughout the week, there were other challenges like sick field staff, reducing the number of teams I could send out. This forced me to quickly rearrange my survey assignments to each group, making sure that the most important areas were covered, but also figuring out which ones I had to let go of. I also found out that areas that seemed accessible from the road, were not without more time to hike and camp. Nonetheless, over two field seasons (returning in 2010 for another four months), I collected over a hundred samples total, enough to analyze for my dissertation.

Before I left my second field season, I spent a week in Libreville, the capital of Gabon, to try to track down the blood samples from the collared elephants so I could add the genetic data to the spatial overlap values between individuals. Back in Missouri, I had emailed everyone even remotely associated with the project, asking them if they knew where the samples were. I couldn't get any straight answers from them; everyone was unsure. But when I showed up at the Wildlife Conservation Society office and asked the receptionist, the samples had been found and they set them aside for me.

I held the plastic Ziploc® bag in my hand in amazement. Part of me couldn't believe these samples were real. Adding them to my study meant my research would be much stronger and I could publish in journals with higher impact. I could see the paper filters with small dabs of maroon-colored blood smeared across them through the plastic along with the names of the elephants I had been studying penciled in. I was so close to packing them in my bags along with my dung samples, but there was one catch. Because elephants are threatened, I needed to fill out a CITES (Convention on International Trade in Endangered Species of Wild Fauna and Flora) permit.

I had everything I needed but one important document, a research permit. And there was a problem; this study occurred before Gabon created such a permitting office for domestic and international research. WCS had conducted the research in conjunction with the Gabonese government, but there were no documents that we could locate that could be used as a research permit for the CITES application because it all happened before the permitting office formed. Disappointed again, there was no way I could legally get the samples out of the country and I was not willing to risk my reputation as a scientist or suffer from any criminal offenses by adding the small Ziplock® bag to my large suitcase full of dung samples (yes, I had a whole suitcase devoted to elephant poop).

Coming back to the US with my suitcase full of poop, I was elated and proud that my field season had been a success. However, I quickly learned that collecting enough fresh samples in the field was one challenge but getting those samples to work in the lab was a new challenge. I first extracted DNA from the dung samples in a private room to prevent contamination to and from the larger, shared lab. Then, I used PCR to amplify the DNA in these extracts just like I did when I volunteered in the flour beetle lab years before. I ran the PCR products on a gel, and if there was elephant DNA present, I sent the samples off to the university's sequencer.

Non-invasive sources like dung and hair do not provide nearly as much high-quality DNA as blood or tissue samples do. When an elephant poops, cells slough off from the elephant's intestinal lining so you're not really sure how many cells you are getting and how viable those cells are. Sun, rain, and time can all quickly degrade DNA. Some samples were so difficult to amplify, that I ran them over a dozen times adjusting small factors like the amount of DNA extract I used to try to get them to work. Some I really needed for my observation networks, for example, if it was one where I was sure the dung came from that elephant. But no matter how hard I tried, I couldn't get them all to work. On top of this, I also had to run all samples that did amplify at least two to three times anyway to verify the genetic patterns I saw.

Graduate school and research were also nothing like I expected it to be. I imagined it would be more like an apprenticeship, where I worked alongside my advisor in the lab, following along with the various protocols she would teach me. In reality, my advisor was never in the lab doing lab work, but instead in her office writing large National Science Foundation grants to try to fund our research, prepping for and teaching undergraduate classes, and writing scientific papers of her previous and ongoing research. This was normal, and my professor was actually more "hands on" than most professors. I've heard stories from colleagues of professors traveling for months at a time and therefore not physically accessible to the student. While she taught me how to do PCRs and run gels in the lab initially, along with some genetic analysis programs, much of my research was different than hers and I had to learn most methods of analysis completely on my own. For any problems in the lab, I first consulted with other labmates to troubleshoot and only my advisor if I couldn't fix it on my own.

Additionally, research moved fast and was complex, especially within ecological genetics (research on genes), and this research was quickly moving towards ecological genomics (research on entire genomes). Prior to graduate school, I thought there were a limited number of ways to analyze data, and that there were standard methods that most scientists used according to the different types of data they had. I envisioned it more like my data analysis in Kenya, where I knew what type of analysis I was going to do, entered the numbers into the program, and point-and-clicked my way through analysis. I didn't realize how many options were out there and how many decisions I would have to be making about analysis. When I interned, I didn't realize how much thought had to be put into understanding what data analysis/program you were going to use prior to analyzing data. Your choice depended on the assumptions of your data and you had to have reasoning behind choosing that method in addition to interpreting the results.

In internships, I was always instructed on what to do. But now I was training to be the expert, conducting research that had never been done before and I had to make the decisions. In genetics, new analysis programs were constantly being developed to analyze data and it was hard to keep up with it all. It quickly became confusing and overwhelming for me and I often questioned if I was making the right decisions. In math class in school, there was always a right answer that indicated if you solved the problem correctly, but in graduate school, there was no known right answer. Even if you arrived at an answer, it didn't necessarily mean that it was "right."

Genetic research was even less confusing compared to statistical modeling, the analyses that most wildlife biologists use. At the time, students were transitioning from the program SAS® to a new free program, R®. SAS® did require writing in some code, but there were less options available in the software, making it a more straightforward approach to data analysis. But R® was another story, essentially a black box for statistics where you could write code to perform whatever statistics you wanted. There were no point-and-click options. This was nothing like the statistics I did in Kenya. Statistics alone was intimidating for me as our school didn't offer helpful courses relevant to our research, and now I had to learn how to write code too.

When I first started graduate school, we were all just learning R®. Now to be a wildlife biologist conducting research in graduate school, you should already come in knowing R® and even how to run models. Such was the case with students in my last lab. This was the important statistics part that my professor in Kenya warned me about, but advanced computing power had really accelerated things. Students in my last lab are even writing code to create entirely new models for analyzing data.

The work and speed of the research combined with all of the analytical choices I had often felt like too much. I was constantly overwhelmed with work and I began to doubt if I had it in me to complete my dissertation. At this point in my career, I really didn't feel good or smart enough.

Everything seemed so much easier for the other students in the department. Their samples all worked better than mine, statistics seemed to come much faster to them, and some of them already had an impressive list of publications. Quite honestly, I was on the verge of quitting.

I was enrolled in a Ph.D. program, so I could easily change to a master's degree and leave with having earned that, so it was not a complete loss of my time. But when I estimated the time I would still need to spend to finish a master's degree, I figured I might as well finish the Ph.D. I had been working on all of my chapters at the same time and data from some parts of my project were involved in other parts. I had already invested so much time and energy into my dissertation and I also truly loved my research subject. It was like my baby and I wanted to see it all come to fruition. I kept my head above the water as best as I could and "just kept swimming" like Dory advised in the movie *Finding Nemo*. It took a lot of sessions of the free therapy offered at my university to persevere.

While I was processing all of my samples, analyzing my data, and starting to write up my dissertation, my fourth year ended, meaning my fellowship was over. For my final years, I had to be a teaching assistant (TA) to a class. I thought I was going to be a wonderful teacher; I was friendly, approachable, bubbly, and didn't use a lot of jargon, but I wasn't. To my surprise, I also hated it, at least the classes I taught. I actually didn't mind teaching general ecology, but it was a lot of work to grade 25 lab reports consisting of several pages every week in addition to their research paper assignments.

In biology for non-majors, the students were required to take a class to fulfill their mandatory science prerequisite to graduate and therefore didn't want to be there. I couldn't blame them. I thought the content was not that important to teach to students as their last course in science. For instance, I was close to finishing my Ph.D. in biology, yet I had to relearn some of the content because I had learned it ten years prior when I was in college. I did my best, but when the students didn't do well on tests, I got blamed by the course instructor. Those students didn't forget their grades or how much they disliked the lab at the end, and I got some pretty harsh evaluations that were difficult to take even though I had no control over the course content.

The following year instead of being a TA again, I applied to be part of a new initiative at our school, the National Science Foundation's GK-12 program. Instead of teaching a college laboratory course, graduate students were paired with K-12 teachers to work alongside them in implementing our research in their classrooms. Our program also required the teachers' students to write "mini grants" to fund equipment to conduct a research project on their school grounds.

I loved working in this program. I was paired with a 5th grade teacher in a rural school outside of Columbia, MO. The emotions I felt from visiting this classroom was in huge contrast to being in academia. The children were so excited to have a scientist in their classroom and had endless questions. During a time when I doubted myself and constantly felt stupid, these kids made me feel like I was one of the smartest and most important people in the world. It was a time when I could forget about my samples not working, lamenting over my broken code, or second guessing myself if I was conducting all of my analyses correctly.

Because our school was so rural, I encouraged the students to purchase a camera trap to study the wildlife around their campus for the mini grant. We wrote up the proposal, were awarded several camera traps, and placed them on various parts of the school grounds. Although the school was rural, it didn't have a lot of habitat, just a small prairie and patch of woods. I wasn't sure we would even get anything that interesting. But when we got the memory cards, we were surprised by all of the animals we photographed: foxes, coyotes, raccoons, deer, and even a bobcat.

Going outside with the students to set up the camera traps and check the photos was so much fun. The students taught me to be curious and excited again about the natural world. I loved working in the classroom so much that I even thought about becoming a teacher at one point. At the end of the year, they gave me a stuffed prairie dog doll, their school mascot. Working with these children revitalized my energy and motivation to finish my dissertation.

In 2012, I turned in my dissertation and graduated that December; my Ph.D. took me six and a half years to finish. Completing my Ph.D. was certainly challenging on its own, but I also experienced major life changes during graduate school that added to the stress and took an additional emotional toll on me. I was forced to confront my mother's declining health. She was slowly, but progressively getting worse from the terminal cancer she developed nine years prior. As a result, I decided to get married to my boyfriend, something I would have normally never done during graduate school because of how time consuming it can be, but I was worried if I waited until after I graduated, it could be too late for my mother to attend. I got married three weeks after I returned from my second field season. If you can avoid it, never plan a wedding while writing a dissertation!

Towards the end of graduate school, I was mentally and emotionally exhausted and did not have the time or energy to invest in looking for jobs or postdoc opportunities. I was only focused on finishing and surviving. My department was not willing to fund me any longer and I had to get my dissertation done. Luckily, one of my committee members had leftover money from a grant and offered me a temporary postdoc to finish some of his research immediately after I graduated. I was grateful for this short nine-month postdoc to give me some additional experience and the time to apply for permanent positions while still earning an income.

In January 2013, I began my postdoc dabbling in disease ecology in the building across the parking lot from where I did my dissertation. My responsibilities included writing a paper on how the distribution of food affected patterns of relatedness and home ranges in raccoons and to carry out a small, but interesting, and slightly crazy, research project. This study investigated "disgust" in raccoons. We set up experiments to determine if raccoons avoided the carcasses of other raccoons at a food source due to higher risk of perceived disease transmission.

This research involved some bizarre data collection on my part. I first established food piles to attract raccoons by dumping big bags of dog food in the research area and affixing camera traps to trees to monitor them. Once the raccoons got used to the food, I placed objects/carcasses at the sites: a stuffed animal (the control), a dead squirrel, and a dead raccoon. I did not kill any animals, but instead, had to find carcasses. Squirrels were easy to find because they frequently are run over by vehicles in neighborhoods. However, finding raccoon carcasses was difficult. Most were on busy highways and too close to traffic to collect safely.

While I conducted this raccoon research, I began my job search. I was most interested in positions that had a direct conservation application and searched for all different types of jobs in many different categories of workplaces across the country. Having finished my Ph.D., I felt a huge sense of relief and that the most difficult times were behind me. I thought my Ph.D. was the key to getting me the job of my dreams. This is what I worked so hard for. As I scanned the job boards on a daily basis, it was exciting to imagine myself moving to places like California or Florida to finally start my professional career. I would finally get a well-paid salary, settle down in a house, and start my new life.

A normal next step for many Ph.D. graduates is to apply for postdoc positions like the one I had, temporary research positions usually lasting between one to three years after your Ph.D. Most jobs outside of academia do not require postdoc experience and I already had some, albeit short, with the nine months after I graduated. I did not want to apply for any postdocs; I wanted a permanent job because, quite honestly, I was tired of moving and ready to settle down. In the past ten years, I moved to eight different apartments/homes not even counting Buffalo when my internships didn't line up.

My husband and I also ended up adopting six rescue pets (I told you I love animals) requiring us to purchase a house in Missouri. But when I found postdoc opportunities I was interested in, I applied to them if they were in larger cities because I thought I could likely find a job there afterward. Although I was not picky about what organization I worked for, I was selective about the geographic location I would live in and was not willing to just move anywhere. I desired a city larger than Columbia, MO and after living in Buffalo, NY most of my life, I knew I wanted some place with no or a very short winter.

Given my flexibility in the exact type of position I was seeking and willingness to work at different types of workplaces (e.g. government, nonprofit, consulting), I assumed I would have my choice of jobs around the country. I started the job search off enthusiastically and picky, only applying for the jobs I wanted most. But over time, my confidence started to fade when I didn't receive interviews. I was taught that job advertisements were a wish list and to apply to jobs where I met at least 70% of the requirements. I then started applying for more general wildlife positions at state agencies, the federal government, nonprofits, and consulting firms. From my days of researching how to write the perfect cover letter and resume, I knew mine were strong, but I still wasn't getting any calls back. What was wrong?

I describe all of this in detail in the next chapter, but in addition to a number of permanent jobs, I ended up applying to three postdocs because they seemed like great opportunities for me and very much in line with my research interests and experiences. For all of the applications I sent out, I was invited to interview for four jobs, three of which were the postdocs I applied for.

Finally, I was offered and accepted a long postdoc, five years at the North Carolina Museum of Natural Sciences (NCMNS). Although I didn't want to take another temporary position, this postdoc really seemed like a perfect fit for me. It involved incorporating camera trap research into middle school classrooms, similar to what I did with the GK-12 program in Missouri. I also greatly admired the scientist I would be working with, Raleigh, NC was a desirable place to live, and given Raleigh was part of the "research triangle" of the US, I was confident I would be able to find a job there afterward.

In January 2014, I moved to Raleigh and began my five-year postdoc with eMammal, a citizen science camera trapping project. In this postdoc, my primary focus was to collaborate with three sets of three different teachers over three years (lots of threes!) to create lesson plans based on my research for teachers to use during the school day. A key aspect of the program was that the data students collected had to be real, useful, and high quality for scientists (me) to publish in peer-reviewed journals. This presented a challenge because I had to think of and design research questions that were scientifically sound, included children, could be done in a short time period during the school day, and align with North Carolina's teaching standards.

In spite of these challenges, I fell in love with this postdoc. My time at NCMNS reminded me of my internship at Disney World and we were even "on display" behind a glass wall for museum guests to watch us as we worked. However, I did have some minor struggles in the beginning and flashbacks of graduate school. Because I was trained in genetics, I didn't know how to analyze camera trap data and immediately felt behind and even stupid at times.

I was also assigned a study on deer vigilance with a large camera trap data set of thousands of deer photos across 33 protected areas. I was under a deadline and felt the pressure to finish, but it was taking me a long time to score all of the deer photos (recording if their heads were up or down), even with the help of undergrads. But those feelings were short-lived overall in this postdoc, and for a good part of it, I felt relaxed, happy, and even took weekends off, something I rarely did in graduate school.

It's difficult for me to sum up all of my experiences working with eMammal in a few short paragraphs. I learned so much and received many unexpected opportunities. Right as I began, we were awarded a grant to conduct our student-led camera trap research not only in North Carolina, but also Mexico and India. As a result of this grant, I traveled to these two countries to train teachers in running camera traps.

In 2015, I was invited to participate in a once-in-a-lifetime expedition. Led by the Smithsonian, we recreated Teddy Roosevelt's original research expedition in Kenya following his presidency in the early 1900s. I spent six weeks camping on Mount Kenya setting camera traps from the bamboo forests at the lower elevations all the way to the barren, rocky top of the mountain.

*In the bamboo forests of Mount Kenya setting up camera traps.*

I now serve as the co-Chair of the Communications and Outreach committee of Wildlife Insights, an initiative composed of seven of the largest science and conservation organizations in the world and the tech giant Google. We are creating the largest database of camera trap images, automated analytics for conservation and wildlife managers, and developing cutting-edge technology like AI to automatically identify species in photos, greatly reducing the time it takes for scientists to process data.

In addition to camera trap research, I also initiated social science projects investigating how camera trap images impact children's attitudes towards local species and if citizen science can form stronger bonds between people and nature. Working at the museum, my science communication skills flourished, and I gained a multitude of experiences in public speaking and creating videos. One of the videos we created on family of red foxes living in suburban Raleigh has reached nearly one million views on YouTube.

My eMammal postdoc though, solidified for me, the importance of understanding statistics and the emphasis of data analysis in wildlife biology. Here, it became even more prominent than it had been in graduate school. A single camera trap could generate thousands of images, multiply that by the dozens of camera traps you need to run a study and the data quickly pile up, requiring more sophisticated and complex statistical modeling for data analysis.

Our lab conducted large-scale research projects. When I arrived, students were analyzing hundreds of camera trap deployments across 33 protected areas in six states in the US. When I left, students were analyzing camera trap data from thousands of deployments across all 100 counties in North Carolina for a state-wide project. Camera trap technology and citizen science allowed for data collection on a massive scale.

In this lab, the students came in already prolific in R® and with prior modeling skills. Here I was with a Ph.D., yet my data analysis skills were far behind them. Students in our lab would joke about writing terrible, inefficient R® code that still led to functioning models, meanwhile I struggled with writing code at all. Week after week in lab meeting, we read papers on the latest models and discussed complex ways to analyze data, such as integrating data from multiple data sources (e.g. camera trap, human observations, telemetry data). The focus was almost always on data analysis. Dealing with the patchier student-collected data though, I was exempt from conducting such sophisticated models.

I loved my postdoc so much and was getting such valuable experience from it that I stayed the entire duration of five years and even had a few months after as an extension. As you'll find out, I also had difficulty securing a permanent job elsewhere in Raleigh. Over those years, I successfully implemented eMammal in over 70 schools across five countries. My research proved that children in classrooms could collect high-quality camera trap data for real scientific studies.

This blend of research, outreach, and science education really made this job a perfect fit for me. The only problem was that it wasn't a real job. It was funded by a grant and therefore would not exist after the funding ran out. Despite my best efforts of working with different state agencies and writing million-dollar grants, I could not raise the funds to extend this position or turn it into a permanent one.

That brings us to now. I know I provided a lot of details about my path, but as you continue reading, you'll find that understanding where I came from, the skills I acquired, and the experiences that I participated in will give you context in understanding my advice. I also hope the stories provided will offer you some perspective on the transitions I had to go through, how expectations didn't always match reality, and some insight on what this career is really like. In the next chapter, I share with you how my experiences played out in the job market. I list all of the jobs that I applied for, which ones I was asked to interview for, and why I was not offered the position.

# 4 ON THE JOB MARKET

*"In this case, it simply came down to finding a candidate that matched our needs almost exactly, and we were fortunate in that regard."*

-Email response from a job I interviewed for that I did not get

### Post Ph.D. Search

Back tracking in time now, we are first going to go back to January 2013, when I just graduated with my Ph.D., eager and excited to find my perfect job. In this chapter, I'm going to revisit my job applications from that period of time. First, I'll review all of my successful job applications by describing the positions, the process, and (spoiler alert) why I wasn't offered the position. At the end, I've provided a list of jobs that I applied for where I never made it past the initial application stage so you can get an idea as to what types of jobs I was not competitive for given my experience. Unfortunately, I didn't keep the cover letters of all of the jobs that I applied for in the first phase of job searching except for a regional ecologist position at *The Nature Conservancy* in Sacramento, California, and therefore the list only reflects the jobs that I applied for after starting my postdoc at eMammal during my second phase of job searching.

When starting my job search, I once again, used all of the knowledge I learned on monster.com years ago to craft the perfect cover letters and resume. I researched the organizations beforehand and sent out dozens of job applications, but I was invited to interview for only four positions. The only problem was that all but one of them were temporary, postdoc positions.

My first interview was for a postdoc position in the Evolutionary Biology and Conservation Science department at the University of New Orleans. In this postdoc, I would have studied Gabonese mammals using non-invasive genetics. It also involved teaching a course to Gabonese students on this subject in Gabon. Given my experience working in Gabon and in genetics for my Ph.D., it seemed like a perfect fit for me.

However, there was one big problem. I was nervous about having to teach a course in Gabon. Teaching ecological genetics is challenging enough in English, let alone in French, the national language of the country. Despite living in Gabon for a total of eight months for my Ph.D. research, I was terrible at learning new languages and I never became fluent in French. While I included that I knew French in the application, I knew this was my biggest weakness.

The professor called me for a phone interview, and it was going quite well until one of the last questions. She explained that part of this job was to teach Gabonese students how to use non-invasive genetic methods and asked me how comfortable I would be doing this, except she asked me this question in French. At that moment though, I had no idea what she was saying. My heart sank. She spoke too fast for me to understand and I didn't recognize enough words. I then had to confess that my French wasn't good enough to understand the question, resulting in her explaining it all to me in English. Given that they were looking for someone to start right away, I knew I wasn't going to get it. I was right, and I didn't get the job for that reason.

The next position I interviewed for was also a postdoc at Disney's Animal Kingdom and in the same endocrinology lab that I worked in as an intern. Deep down inside, I think I knew I didn't have the best chance at getting this job, but I couldn't resist applying because I loved my internship at the Wildlife Tracking Center. I thought if I could get this job at Disney, that perhaps I could find a permanent job there afterward.

Like my internship, the position focused on endocrine research, but obviously at a much more advanced level as this job required a Ph.D. Although I could easily relearn the laboratory protocols, especially since lab research was such a large component of my Ph.D., ultimately my degree was not related to endocrinology. Therefore, I lacked the background knowledge necessary to conduct research at the postdoc level. Given that postdocs are temporary, there is not enough time for one to spend catching up and most employers want someone who can come and hit the ground running. My phone interview went well, and I even interviewed with some of the people I used to work for. It was great to catch up with them, but afterward, I even felt like this postdoc wasn't right for me. I understood their decision that I was not the best candidate and I also agreed with them; it would have been very challenging for me to switch gears so fast.

Luckily, I was finally invited to interview for a permanent position, and it was one that sounded perfect for me: Assistant Curator of Conservation and Research at the North Carolina Zoo. The position entailed conducting behavioral research on the captive animals at the zoo and my own research internationally. The zoo was especially interested in research in Africa, as they already had an educational program for K-12 students in Uganda.

The NC Zoo flew me from Missouri to Greensboro, NC for an in-person interview. I was instructed to prepare a presentation on my research to the search committee in addition to my interview. On the day of, I met some of the researchers and toured the zoo. It really seemed so perfect for me. The only problem was the location. The NC Zoo is located in a small town of roughly 25,000 people. Columbia, MO, which had over 100,000 people was already too small for me. No matter how much I loved my job, I knew I would not be happy living in Asheboro. My other choices for residency were either Greensboro, which was a 30-45-minute commute, or Raleigh, which was an hour commute, but a much nicer city than Greensboro. My husband started to look at his job prospects in Greensboro, but nothing showed up for him. It looked like Greensboro wasn't working out for us, but an hour commute was also a deal breaker for me.

At the same time as I waited to hear back about the zoo position, I interviewed for another postdoc. This position was also in North Carolina, but at the Museum of Natural Sciences in Raleigh and with a scientist I admired. It was for the eMammal project, which involved working with volunteers in running camera traps to study mammals in human-developed landscapes like people's backyards, cemeteries, golf courses, and schools. I would implement the research into classrooms by collaborating with K-12 teachers, which I had already been doing in my final years as a graduate student in the National Science Foundation's GK-12 program.

I interviewed with a large team of people from the museum and the Smithsonian, partners of eMammal, over Google Hangout. The interview went well, but I wasn't confident that I would get it. I struggled thinking of exact research questions to ask with data collected by kids in schools. Research had to be standardized and there were so many factors to consider when working in a school system, but to my surprise and delight, I was offered the job in the next few days.

The eMammal team wanted a response within the next week because they wanted someone to start soon, but I still hadn't heard back from the NC Zoo on the assistant curator position. If eMammal had been a permanent position, the decision would have been easy. But indeed, it was a temporary postdoc, while the NC Zoo job was permanent. In the end, I decided it was better to have a real offer, albeit temporary, than one I thought I had, and accepted the eMammal position. Although I really wanted the NC Zoo job, it was too logistically challenging to make it work for my husband and me. I was also certain I could find a permanent job in Raleigh, NC within the next five years. Raleigh was a large city made up of half a million people and combined with Durham and Chapel Hill, it made up one of the corners of the "research triangle." This was a good decision because I ended up not getting the zoo job and I am not sure why.

Once I started the eMammal postdoc in January of 2014, I was technically in my second phase of job searching, but I was selective in the jobs that I applied for. This time I really wanted to stay in Raleigh, so I almost always excluded jobs that were located elsewhere. I knew this would decrease my chances of securing a permanent job but staying in Raleigh was really important to me for several reasons. I suffer from an autoimmune thyroid disorder and at the time, undiagnosed chronic fatigue. Because of my fatigue, moving was physically and emotionally taxing for me. In Raleigh, I also started to find doctors that were really helping me.

Additionally, once you have a significant other, you have to consider your partner's career in addition to yours. People often call this the "two body" problem. Technically my husband could find work in another location, but just because he could, didn't necessarily mean that he wanted to. It also wasn't looking good on his resume to only stay at companies for a year or so. At this point, we had already moved for my job twice, and we needed to consider his career. The "two body" problem is especially difficult when both people are scientists, and you both have to find jobs in the same city.

I really loved my job at eMammal. It was my dream job except for the fact that it wasn't permanent. Through eMammal, I was offered amazing research experiences, developed collaborations across the world, and worked alongside top-level research scientists. Whenever a job was posted that I qualified for, I carefully weighed whether it was worth giving up these experiences for, as I still had more years of funding in my postdoc. With eMammal, I was working on cutting-edge projects and I never knew what would happen in the future.

I really loved my job at eMammal. It was my dream job except for the fact that it wasn't permanent. Through eMammal, I was offered amazing research experiences, developed collaborations across the world, and worked alongside top-level research scientists. Whenever a job was posted that I qualified for, I carefully considered whether it was worth giving up these experiences for as I still had more years available to me in my postdoc. With eMammal, I was working on cutting-edge projects and I never knew what would happen in the future.

Additionally, applying for jobs is a lot of work and actually takes time away from your research. You need to craft a cover letter tailored towards each position, and as you will find out, the jobs that I applied for were very different from one another, which made it hard to repurpose letters. Many positions required a research statement, teaching philosophy, and/or writing sample. Once you get an interview, you will likely have to give a presentation or draft a mock proposal.

I learned from phase one of my job search that I was only a competitive candidate for jobs when I met at least 90% of the requirements and desired qualifications. This is in contrast to what I learned from job websites and other experts. I was often told that position advertisements are "wish lists" and that employers know they won't get candidates that meet all of the requirements. If you meet ~60-70% of the qualifications, you should apply. However, as I learned in my first phase of searching for jobs in wildlife biology, this was not true. Therefore, in phase two of my job search, I didn't invest the time applying to jobs where I didn't meet a vast majority of the desired criteria.

**Post Postdoc Search**

My real dream job opened up very shortly after I started the eMammal postdoc and this time it was permanent. The position was for the Assistant Director of the Biodiversity Lab, where I was already working, and it was an entry-level, post Ph.D. job. If I was offered the position, I would be conducting my own research on anything I wanted as long as it involved the life sciences. By then, I already loved working at the museum and with eMammal. I developed strong relationships with other scientists and truly enjoyed the work environment. I spent weeks perfecting my application, and it paid off.

The museum invited me to interview and I poured my whole heart into prepping for the interview, which included two talks, one for the general public and one for an academic audience. I spent weeks researching the museum, planning, and rehearsing my talks. Prepping for the job and giving the talks only made me more excited about the possibility of this position. With this job, I could return to studying forest elephants too, and have a long-term research project in Gabon. Despite my best effort, I didn't get the job.

Instead, the position was offered to a scientist who had ten years more of experience than me. Because of these ten additional years, she was a better scientist. She wrote more publications and was awarded more grants than me, qualifications that search committees usually care about most. There was no way I could compete, and as a result, I was choice number two. I would be lying if I said I wasn't heartbroken at the time.

In November 2017, I interviewed with the North Carolina Wildlife Resources Commission (NCWRC) for the Diversity Outreach Specialist position. The successful candidate would develop educational and outreach programs targeted towards underrepresented communities to increase awareness of the NCWRC and increase participation in wildlife-related activities. From our eMammal state-wide camera trapping program, I was already working with two NCWRC staff on the search committee and they were both supervisors for the advertised job.

I spent weeks researching the NCWRC and brainstorming creative educational and outreach program ideas. I was asked to prepare an example of a program I would implement and wrote about connecting people, especially kids, to urban nature through citizen science like I had learned in eMammal. Despite these connections and my enthusiasm, I did not get the job. Again, someone else more qualified was offered the position. She had years of experience developing similar programs at another institution and spoke Spanish fluently, making it easier for the NCWRC to reach Hispanic communities.

I was invited for my next phone interview in early 2018 for the United States Executive Director position at the nonprofit Friends of Bonobos. Again, this position was ideal for me. It was all about conservation, which was always my driving motivation in pursuing a wildlife biology career. Friends of Bonobos supports *Lola Ya Bonobo*, a sanctuary in the Democratic Republic of Congo that rescues confiscated bonobos from the illegal pet and bushmeat trade. They conserve the species through rehabilitation and reintroductions of bonobos, and education and outreach to the local community, many of whom had never seen a bonobo before.

I thought I had a good chance at getting this job as working at the museum made me an excellent public speaker, I worked in the Congo basin for my Ph.D., and conducted animal behavior research on a socially complex, highly intelligent species. However, I had no formal fundraising experience. I was confident I would be able to fundraise, but I had no demonstrated experience to vouch for this. Therefore, the job was offered to someone with fundraising experience.

I was invited for two rounds of interviews for this job with two mostly different search committees. For this position, I had to prepare a proposal explaining how I would go about conducting a social science study on an endangered species. I was devastated when I noticed I wrote "North California" instead of North Carolina in the title of my proposal, but this was overlooked. I was told later that I was unsuccessful in convincing everyone on the search committee that I could conduct social science research independently. In my interviews, I frequently used the word "we" to describe how collaborative my research was with people from other institutions in the area. I thought this demonstrated how well I worked with teams and on large projects, but this combined with the fact that I didn't have a degree in social science, apparently made some people doubt that I could carry out the research independently. I failed to convey that I wrote the surveys, administered them to students in classrooms, collected, and analyzed all of the data myself.

This job was actually reposted weeks later. It was a failed search for the NCWRC, and they reopened the position. One person at the NCWRC actually encouraged me to reapply, suggesting that I just needed to be more convincing and that I still had a good chance. Ultimately, I decided not to because after learning more about the research at the in-person interviews, I was not as interested in some of the projects (e.g., conducting studies on boat access points).

By now, my postdoc was coming to a close and I started to panic and feel hopeless. What would I do if I didn't have a job? I couldn't believe it had been so hard for me. I applied to so many positions where I thought "this was it," but I didn't even get an interview.

My next job interview was for a public communication specialist within the Applied Ecology department at North Carolina State University in early 2019. One of the professors on the search committee was the lead on the grant I was hired under and I technically worked under him, even though my position was at the museum. This position was a mixture of different jobs, but the main focus was on communicating the research findings of the Applied Ecology department to the public. Social media was a big component of the job, but fundraising, coalition-building, and departmental research organization also made up major responsibilities of the position. When I interviewed for this job, the professor interviewing told me that I had the C.V. (academic resume) of an assistant professor and asked me why I would want a job like this. I explained my love for science communication and ecology, but I didn't get the job and I'm not sure why. I suspected it might have been because I was overqualified for the position.

To my surprise, in 2019, the Assistant Director position of the Biodiversity Lab opened up again. This was the position that I was runner up for in 2014 at the NCMNS. At first, I was excited, but when I read the posted job description, I was disappointed to see changes made to the job that would make me a less competitive candidate.

The position advertised specifically for a geneticist and now paid at least $20,000 more in salary, but upwards of $70,000 more, than the posting in 2014. Although I conducted genetic research for my Ph.D., I had not done any new research using genetics for almost six years at that time. Genetic research changes very fast and I knew that I would have a lot of making up to do in terms of knowledge and learning new methods. Also, because I was conducting research under the Director of the Biodiversity lab and at the museum for the last six years, my research was now very similar to his. I was told previously that I was not competitive for a professor position I applied for at NC State University for this exact reason. He had a joint position and was a professor there too. I knew I had the same disadvantage at the museum.

After long deliberation, changing my mind back and forth several times, and having even started writing a research statement, I decided not to apply for this position. When I started writing the research statement, I struggled to come up with project ideas and wasn't even that excited about the ideas I came up with. I discussed my research statement with a friend and colleague for advice, and she reminded me that I was never really happy doing genetic research. I knew she was right. I was never driven by genetic research itself. I was mostly interested in using it as a tool to answer ecological or conservation questions. And even then, I still didn't enjoy the laboratory aspect of it. I enjoyed camera traps much more. Day in and day out, I knew I wouldn't be happy mentoring students, teaching classes, and conducting research that I was not passionate about.

I applied for one more job recently on a whim, even while writing this book. It was far outside of Raleigh at the Cary Institute of Ecosystem Studies in Millbrook, New York and aligned very closely to my interests in K-12 education and citizen science. It was an Ecology Education Program Leader and very similar to my research and outreach with eMammal. I would be working with K-12 teachers on citizen science and outreach programs for the Cary Institute. I knew I was competitive for the job and was called for an interview the day after I applied. During the phone interview, the head of the search committee disclosed the salary, something that they did not post in the advertisement. The job didn't pay enough to make financial sense for my husband and I to move away from Raleigh and I withdrew my application.

Those were the jobs that I received interviews for. Below is a list of jobs that I applied for where I was outright rejected:

1. Assistant professor position researching the ecology and conservation of large mammals, University of Florida, October 2015
2. Assistant professor position in Natural or Social Sciences as part of the Leadership in Public Science cluster, North Carolina State University, October 2015
3. Curator of Education, NC Zoo, December 2016
4. Technical Writer, National Institute of Health, August 2016
5. Vice President of Research, Half-Earth Project, September 2017
6. Assistant professor studying urban ecology in the Applied Ecology department, North Carolina State University, September 2017
7. North Carolina Field Campaigner, Center for Biological Diversity, November 2017
8. Mammologist, North Carolina Wildlife Resources Commission, Jan 2018
9. Public Communications Specialist, Water Resources Research Institute of the UNC system, January 2018
10. Blog Managing Editor position, Red Hat, February 2018
11. Assistant professor in the Biology department, Shaw University, August 2018
12. Education & Communications Coordinator position for the Center for Galapagos Studies, University of North Carolina at Chapel Hill, September 2018. *Note: This job ended up being cancelled by the employer.*
13. STEM in the Park Program Manager, Research Triangle Park,

January 2019
14. Director of STEM Learning, Museum of Life and Science, June 2019
15. Lecturer position, Department of Applied Ecology, North Carolina State University, June 2019

Quite honestly, most of the rejections were shocking to me. I learned in graduate school that academic positions were competitive. According to a study in 2020 by Jeremey Fox[1], Professor at the University of Calgary and creator of the *Dynamic Ecology* blog, the average tenure track ecology job seeker applied to 15 different positions every year, but with a range of up to 100 positions. But I never learned that non-academic jobs were competitive too.

Maybe you are thinking it is just me or that I had bad luck. But social media, especially Twitter, is rife with comments about how competitive the job market is (some of which are included in the course notes). Shortly after graduating, I remember my friend, who was a year ahead of me in the process, had a difficult time getting a job. My friends and I helped her troubleshoot the situation. We asked if she was applying to jobs she was qualified for and if she tailored her cover letters to each position. She said yes to every question we asked. At the time, I couldn't understand how she could send out so many applications, but not get a permanent job, and thought maybe she wrote her cover letters in haste. Now that I've gone through the experience myself, I understand it well.

Almost every job rejection was painful, but some hurt more than others. At one low point, I became extremely frustrated when I was rejected for a women in STEM (science, technology, engineering, mathematics) fellowship focused on outreach to middle school girls. Having worked with eMammal integrating our research into K-12 classrooms, I had literally been doing that for the past five years. I felt defeated. If I couldn't get a fellowship I was extremely competitive for, how was I ever going to get a permanent job?

Because I didn't know how competitive jobs would be once I graduated, I didn't take the time to imagine what types of jobs specifically I could get with my experience. Honestly, I never knew where I even wanted to go in my career, I just knew I wanted a job that involved research and conservation. My interests spanned across different taxa, ecosystems, and geographic regions. I thought I would be able to choose amongst a wide variety of jobs post-graduation and the right job would speak to me. I had failed to realize that I would not be competitive for those jobs even though I had a Ph.D.

I originally chose to study forest elephants because I truly loved them. Elephants completely fascinated me, and I could not believe how little

---

[1] Fox, J. A Data-Based Guide to the North American Ecology Faculty Job Market. *Ecology*. 101: e01624

research had been done on this species. I loved working in Africa, and I knew I would not get bored studying them for more than five years. My research questions were interesting and something worthy of study. Unfortunately, I found the skills that I learned from my dissertation were not appropriate for the types of jobs that I wanted and could easily be transferred over.

My work in understanding forest elephants' patterns of relatedness and social networks gave me expertise in non-invasive genetics and behavioral ecology. Before I started my dissertation research, I should have looked at the job boards to see what kinds of jobs I could get with those skills, working with that species (or a similar one), in that area of the world, and how frequently those types of jobs show up. In graduate school, I got the impression your study species and area of geographic region didn't matter much, and it was more about your skill set and having completed a Ph.D. However, when I applied to jobs in other areas of the world, they desired scientists familiar with the taxa and ecosystem for the position.

My dissertation research took place in Central Africa and on forest elephants, therefore making me most knowledgeable about elephants and that area of the world. There are elephant jobs in Africa, however, I now had my husband and pets to think about. I am also close to my family, so I didn't want to live permanently that far away. Going to Gabon was great for fieldwork, but I didn't think far enough ahead in my career to realize that most elephant researchers or conservation biologists conducting research in Africa would live there. This sounds crazy in retrospect, but I visualized having a job where I could live in a city in the US, and travel to field sites for work. You can do that as a professor at a university, but it's rare in other kinds of workplaces.

Additionally, most international conservation organizations want to hire in-country scientists over expat ones. Conservation is more effective when programs are designed and led by the citizens of that country; they know the people, their cultures, and can speak their language. The local community is more likely to trust this person too. Even if I did aggressively search for jobs in Africa, I would have had a harder time getting a job in Central Africa, where I did my dissertation research. I really struggled with French, the national language of Gabon and many other Central African countries, making me a weaker candidate.

What I wish I had done instead was think about the job I wanted immediately after graduating and ten years after that. Before I chose my dissertation research, I should have looked at government jobs and jobs at nonprofits like WCS and the World Wildlife Fund (WWF) to see what types of positions were available, where they were located, and what I would be qualified for. Most research positions for WCS and WWF are internationally based, while most jobs in the US for these nonprofits are focused on marketing, development, and fundraising. Although I loved my dissertation research and am grateful for my field experience, it best prepared me for a job in academia or zoos. I never wanted a job in academia, and would love to have a zoo job, but they are difficult to obtain.

My skills did transfer over to North American wildlife, but in the positions I applied for, there were probably many candidates who were already trained and did their research in North American ecosystems. If an organization had the option to hire a biologist who worked in Gabon or one who worked in the same ecosystem and location that the position advertised, they are more likely to go with the latter. Given how many qualified candidates there are at all different degree stages, competition is high, and employers are likely going to hire the candidate who is ready to start with the least amount of training.

Although my postdoc was supposed to help me get a job with more experience, the longer I stayed in it, the more similar my research became to my advisor's research. Given he had dual employment at the museum and NC State University, this removed two workplaces that I would not be competitive at because they would not want to have two researchers doing the same types of projects. They would rather have diversity in their program subjects.

In addition to the high rate of rejections, I was also surprised by how few jobs were available that I was competitive for. Five years seemed like plenty of time for me to secure a permanent job, but as I kept my eyes on the email alerts I was subscribed to, I would rarely see wildlife biology research positions at my level of experience. Even outside of Raleigh, many of the jobs advertised were at the very entry-level (e.g., internships, technician positions) or so advanced that you needed at least a decade's worth of experience after your Ph.D. In the end, most of the jobs came from NCWRC, NCMNS, and NC State. I expected there to be more federal positions available such as with the US Fish and Wildlife Service or the US Geological Survey, which hires biologists too.

The greatest lessons that I learned from these rejections was that this field is saturated, or at least specific disciplines within this field, and to be competitive, you pretty much have to match what an employer is looking for. Looking back on my applications, I only received interviews for positions where I was highly qualified. Even in positions where I matched the qualifications exactly, there was always someone else more qualified than me, making them a better fit. I wrote this book so you can learn from my mistakes and make sure you get the exact experience you need to be one of those perfectly suited candidates. Throughout this book, I break down careers in wildlife biology and strategies that will lead you to your perfect position, including jobs tangential to wildlife biology.

# 5 KNOW YOUR DESTINATION

Most aspiring wildlife biologists make the mistake of following their passion, like I did without exploring the ultimate options they will have in job opportunities. I belong to a wildlife jobs networking Facebook group and frequently there are questions asking for advice. Students are always asking what skills they should acquire, what degrees they should get, and what research they should do. My response is always the same. What do you want to eventually do? The answers for all of your questions, from what courses to take and experiences you need, are all in the job boards.

In the last chapter, we learned from my experiences on the job market that applicants who were offered the job were the most qualified for the position. Therefore, my biggest advice for you is to truly understand your final career destination. Or at the very least, understand the various career destination options and what they require. If you fail to do this, you'll miss out on the key skills and experiences that you need to get the job of your dreams. Don't rely solely on advice from your professors or other professionals. Of course, you should be asking others for advice, but keep your eyes on the job boards. The standards in this career are constantly changing and those who have been in the field for a long time may be out of touch with what is needed to get a job today.

Imagine you won free roundtrip tickets to anywhere in the world. What would be the first thing that you do?

What country? What will you be doing? Climbing Mt. Everest or shopping at high-end boutiques in Paris? Do you want to stay at fancy hotels? Or backpack in hostels?

If you pack your bags without knowing where you are going, you could end up with all of the wrong stuff. You can't wear fancy dresses and high heels on a trek up Mt. Everest, and you won't want to wear winter coats and boots in the Bahamas. Without knowing your final destination, you can't be sure you will have the clothes and equipment you need to carry out your trip successfully.

The same is true for careers in wildlife biology. There are lots of final destinations and even more when adding careers tangential to this field. You don't want to spend a lot of time preparing for this career only to find out that you've missed out on some of the key requirements that you need for the job you really want. Once you narrow down what type of wildlife biologist you want to be and where you want to work, you can get the right training you need to be competitive for the job.

There are two reasons why this is important: (1) As mentioned before, the job market is saturated and to be competitive for these jobs you will need to be the closest match to the job description posted in the advertisement and (2) What you think a wildlife biologist is may not match up with reality. It didn't for me, and I know this is true for many others because I frequently receive emails from people who are interested in this career but have the misconception that we frequently work in close contact with animals or are always outside. I didn't truly understand what a wildlife biologist did until I went through graduate school.

I know what you are thinking; you don't know exactly where you want to end up in your career. Don't panic. I was in the exact same place as you. But unlike me, you have this book and the accompanying resources in the online course that will act as your North Star to help you figure out where you want to go. It's okay that you don't know exactly what you want to do at this moment, but my goal is that after reading this book, you will be clearer on your final destination. You will also know what basic qualifications you need for different types of jobs if you are still deciding between final destinations.

First, let's go over some of the basics about what wildlife biology is and what wildlife biologists do. The answers may surprise you.

**What is a Wildlife Biologist?**

If you were to do a Google image search for the word "wildlife biologist," you will see mostly photos of people holding animals, especially large, charismatic mammals like big cats. Go ahead and do it now. What did you see?

I've been a wildlife biologist for 17 years now and I've never held a non-captive wild mammal. I know this sounds crazy, but although I did specialize in large mammals, all of my work was non-invasive. I have been, however, in the presence of other biologists holding wild mammals and have held wild animals myself like frogs, toads, and snakes for fun and research.

What I want to stress is that although it is absolutely true that some wildlife biologists touch wild animals, the actual handling is usually a minor part of the job. Let's first go over exactly what a wildlife biologist is.

Biology is from the Greek words "bios," which means life and "ology" denotes a subject of study or interest. Biologists, therefore, study living organisms. They do so through research using the scientific method.

Wildlife refers to wild animals, specifically those that are undomesticated and living in the wild. Management is usually focused on species living in their native range. Wild animals outside of their range are considered invasive species, and scientists almost always try to eliminate these populations. Captive, undomesticated animals are not considered wildlife.

A wildlife biologist, therefore, studies wild animals and often in context of their ecology. Wildlife biologists may research areas like abundance, range, behavior, diet, habitat selection, reproduction, species interactions, community dynamics, and human-wildlife interactions, to name a few.

More recently, the term wildlife has been extended to plants and other organisms, but for wildlife biology as a career, the definition pertains to wild animals. For careers in wildlife biology, wildlife is almost always referred to as vertebrate animals. For example, an entomologist, someone who studies insects, is usually not referred to as a wildlife biologist even though insects are technically wildlife.

In terms of jobs, wildlife biology usually refers to terrestrial wildlife. While many of the concepts in wildlife biology most definitely apply to marine and aquatic species, people who study marine species are usually called marine biologists. Wildlife departments at universities are often split with separate courses for aquatic species in the fisheries portion of the department (e.g., Fisheries & Wildlife department). In job listings, positions researching or managing fish are usually listed as fisheries biologists. I am not an expert in marine biology, so most of my advice throughout this book applies to terrestrial species, especially mammals, my area of expertise. In the course notes, I included links to excellent podcasts run by marine scientists that offer career advice.

You may also hear other "ologist" names that include wildlife, such as mammologists (mammals), ornithologists (birds), herpetologist (amphibians and reptiles) and ichthyologists (fish). These titles can be tricky, though, because the research that someone conducts under these titles will vary according to where they work. A mammologist working in a state agency will conduct research more like a traditional wildlife biologist, whereas a mammologist in a museum is more likely to conduct research related to taxonomy and the evolution of species. They will therefore have different areas of expertise and skill sets.

The wildlife biologists you see holding charismatic mammals in the Google search results are usually researchers that study animal movements. They capture animals temporarily, anesthetize the animal, and attach a tracking device to the animal. The fieldwork associated with this process makes up a small portion of the researcher's daily work, maybe only a few weeks a year, if that. The actual handling time of the animal is very short, less than an hour, as scientists need to minimize stress to the animal.

The tracking devices are expensive, so few animals will be tracked to begin with. The technology has improved significantly, with many trackers now providing data for years. Scientists have also switched from radio telemetry to satellite telemetry, eliminating the need for researchers to go into the field to get readings. Once an animal is tagged with a tracker, you collect data from your office indoors.

There are careers in wildlife biology where you handle animals outside of animal movement research, but those careers are usually with smaller animals. Wildlife biologists working with small mammals, bats, reptiles, amphibians, and birds capture animals for data collection (e.g. measurements, genetic samples, to tag them) and then release them.

*GPS-collared elephant I saw in Loango National Park, Gabon. Researchers received location data from the elephant's collar without having to go to the field.*

A lot of people apply the word wildlife biologist to anyone who works with exotic animals, but this is incorrect. A zookeeper, wildlife rehabilitator, or an animal sanctuary employee are not wildlife biologists because they do not study the animal *in situ* (in the wild). Those careers require different skill sets and knowledge such as animal husbandry and veterinary medicine.

## Differences Between Wildlife Biology and Zoology

Zoology, "the scientific study of the behavior, structure, physiology, classification, and distribution of animals," (definition from the Oxford Dictionary) is similar to wildlife biology, but some nuances make it different. The major difference is that wildlife biology is focused on free-ranging populations of wild animals and often includes management and applied research.

One would be a zoologist but not a wildlife biologist if they studied wild animals in a lab or a captive setting. Scientists at the Duke Lemur Center in

NC study the behavior, cognition, communication, physiology, aging, and genomics of the lemurs at the facility. These scientists would be considered zoologists over wildlife biologists because the lemur population is captive and *ex-situ* (not in their native habitat). The questions that zoologists ask tend to be outside of the ecology of the species they are studying, whereas wildlife biologists tend to focus more on questions relating to their ecology. That being said, if you are a zoologist, you may qualify for some jobs in wildlife biology and vice-versa depending on your experience and the hiring organization.

## What is Research?

Wildlife biologists study wildlife using the scientific method. Quite often in school, even in college, you are taught that biology (or science as a whole) is a series of facts and concepts to memorize. Lots of students think biology is hard because there is so much to memorize: different organisms and the taxonomic groups that they belong to, the stages of cell division and when and where there are 1 or 2 ns, and the Hardy-Weinberg equilibrium equation to name a few. If you did happen to love your biology courses, this does not necessarily mean that you will love being a scientist.

Being a scientist is not about memorization at all. Science is about figuring out answers to questions that have not yet been solved. In biology class, teachers go over the scientific process: forming a question or hypothesis, making predictions, collecting data with a proper protocol in place, analyzing results, interpreting results to see if they match your predictions, refuting/failing to refute your hypothesis, and discussing the results. In school, the laboratory portion of a class is designed for you to carry out this process and learn the scientific method. However, I believe most labs fail to do this.

The labs I participated in or taught have almost always been designed to give you an outcome to a question that has already been studied. It's often on a topic or concept that is important for you to memorize in the course. Frequently, we learned the concept first in the lecture portion of the class, so I already knew what type of result to expect going into the lab. This approach fails to teach you what science is really all about, asking questions when the results are unknown by building off of previous research.

It's easy to come up with a scientific question, hypotheses, and predictions when you know the concept that the lab is designed to show you, or when you know what experimental protocol to follow. As you saw for me, coming up with an original question was hard. The process and successes you experience in a biology lab, and even in professional research settings later on as a field assistant or intern, will not match up to your expectations of being a scientist if you are not the one coming up with the questions.

In biology class, if you study hard and put in the work for the lab, you will get a good grade and succeed. As a research scientist in real life, coming up with questions can be difficult. They need to be interesting, something that someone has never done, but something that you think you would be able to carry out successfully and can afford to do with grant money. Additionally, it is difficult to obtain funding for your research, things always go wrong during data collection, and the results are usually messy, making them difficult to interpret. Being a scientist requires curiosity and creativity, which is in direct contrast to how you are taught science in school, or at least how I was taught. We mostly learned that there was a single right answer and that you had to memorize it.

When people hear the word research, in addition to watching animals, they also think about the research that they did for writing papers in school. Writing papers for class requires a lot of reading and the bringing together of multiple sources of information, most of which were from books and encyclopedias for me growing up, but now are sources on the Internet.

Real scientific research does involve a lot of reading, but it is almost always from peer-reviewed journal articles. Reading a single journal article well takes approximately forty minutes to one hour. It takes skill and practice in knowing how to read scientific articles, and you will need a foundation of knowledge within the field you are reading about. Reading the literature from previous scientific studies conducted allows you to decipher the gaps and/or conflicting outcomes in science and acts as a roadmap to your next research question.

As I learned in graduate school, scientific research is really all about asking questions. If you are a child reading this who wants to become a scientist, keep asking questions! If you are a parent raising kids who want to become scientists, encourage them to ask questions. I believe it is especially important for adults to encourage question-asking because I feel that the educational system in the US can discourage it.

Throughout school, I was constantly told there was one right answer. If I had to ask questions, I thought it reflected on my inability to learn, meaning that if I asked a question, it meant that I didn't understand what was being taught. I was afraid other students would think I was stupid. But rather, I should have been asking questions to push the teacher's knowledge limit and think beyond what I was just told. Girls are especially vulnerable to this because we are encouraged to be polite and not disruptive. Asking questions about the natural world was the hardest thing for me to relearn when I became a graduate student, and I still struggle with it today. This skill will serve you well if you are interested in a career in research, and it pays off to cultivate it from a young age.

Within wildlife biology, there are two broad categories of research that I will refer to throughout this book: applied and theoretical. Let's take a look at the differences between them.

## *Applied Research*

Applied research refers to research where the questions developed have specific conservation and/or wildlife management outcomes. There is an action in mind depending on the results of the research. Nonprofits and government agencies almost exclusively conduct this type of research. They ask and answer specific research questions to obtain more information to help them manage wildlife for conservation and to maintain healthy populations.

For example, when I was in graduate school in Missouri, the state agency that manages wildlife, the Missouri Department of Conservation (MDC), wanted to know more about their seemingly growing black bear population. They funded research in our lab for a student to set up barbed wire fences around attractive lures to non-invasively collect bear hair for genetic analyses. The student used the extracted DNA to estimate how many black bears were in the Missouri population and investigate gene flow between the original bear population and others that were reintroduced to the area. The answers to these questions are useful to the MDC, were provided in presentations and reports to the MDC by the student, and they will use the population estimates generated from the study in bear management.

## *Theoretical Research*

Theoretical research is research that advances knowledge for knowledge's sake. Theoretical research helps scientists collectively better understand species, ecosystems, and community dynamics in general. The research outcomes can absolutely be used by the government and nonprofits for conservation and wildlife management purposes, but typically the research questions are designed independently.

During my eMammal postdoc, I led a study on deer vigilance in relation to human recreation, hiking, and the presence of coyotes. Using a large-scale dataset of deer photos from the eMammal project, we recorded the head position of solitary white-tailed deer across 33 protected areas in the eastern US. We investigated if humans and coyotes could act as apex predators in this ecosystem given that the natural apex predators of deer, grey wolves, and mountain lions, were extirpated from this area.

This study addressed an interesting research question about predator and prey dynamics; however, the outcome of our project was not useful for the management of a specific area or deer population by a state or federal agency. Rather, this information provided general knowledge about how deer respond to the presence of coyotes and hunters across parks. We found no relationship between deer and either coyote abundance or human hunting through increased vigilance, suggesting that deer do not perceive coyotes or hunters as apex predators.

Theoretical research is often conducted by professors at universities but can also be conducted by governments and nonprofit organizations. For the latter two, usually, it is embedded in a larger conservation or management question because theoretical research is often not their highest priority. Additionally, research can be both theoretical and applied. In the bear example I used above for applied research, the student published theoretical papers on the ability of reintroduced populations to restore genetic diversity and the effectiveness of novel genetic markers in analyses compared to traditional methods.

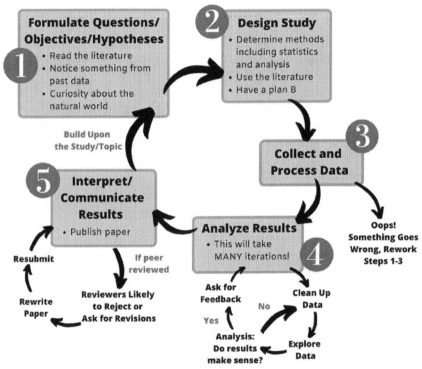

*My recreation of the scientific method from what I learned after graduate school.*

Before we move on to the state of wildlife biology now, I want to take a moment to discuss publishing as it is something that will be mentioned throughout the rest of this book. A big component of research is publishing in peer-reviewed scientific journals. Search committees for jobs often evaluate you as a candidate based on your publication record. If you remember the model of the scientific method, the last step is communication.

Publishing is the most important way in which scientists communicate with one another. If you did an experiment or researched something but never shared the outcome, what would be the point? Science builds off of previous knowledge, and this knowledge is shared in peer-reviewed journals or what scientists collectively call the literature.

Publishing can be extremely important in wildlife biology. For many jobs, especially those in academia, a publication record is required. Typically, the more publications you have, the more competitive you are for a job. However, quantity is not everything, search committees also look at the quality of journals that you publish in and whether you are first author.

Because the job market is saturated, you will need more publications for the same type of job than was needed ten or twenty years ago, and you will need publications for jobs that formerly didn't require any. I know of one professor at an R1 university who got his job with only one publication. Meanwhile, my friends and I applied for university jobs and did not even get interviews with 10-20 publications. In an article in *Nature Research*, a distinguished professor in psychology said "When I started my career almost 30 years ago, a few peer-reviewed publications could secure an academic job...Today...a CV that used to get you a job now makes you competitive for a postdoctoral fellowship."[2]

## Wildlife Biology: Past and Future

The field of wildlife biology has changed not only in higher expectations and more requirements of applicants to secure a job but also in the nature of the work itself. Technology has revolutionized the way we study wildlife and analyze data. It has also allowed science to move at a much faster rate.

The only wildlife biologists I knew growing up were Jane Goodall and Dian Fossey, groundbreaking female scientists who studied chimpanzees and mountain gorillas in the remote forests of Tanzania and Rwanda, respectively. I went into wildlife biology thinking my research would be similar to theirs, and to the researcher I met in Amboseli National Park who studied elephants.

I imagined myself working in a similar environment, observing the natural behaviors of some charismatic species in an exotic, faraway place. I did fulfill this dream during my Ph.D. research; I studied African forest elephants in the forests of Central Africa. I spent eight months of my life watching wild elephants in the field in a national park in Gabon. But after graduation, I quickly realized there are very few permanent jobs where you can do this type of research. Given the advancement of technology and computing powers, I

---

[2] Reinero, D. *Nature Research*. https://socialsciences.nature.com/posts/55118-the-path-to-professorship-by-the-numbers-and-why-mentorship-matters

presume there will be even fewer and fewer jobs like this in the future.

Scientists like Jane Goodall, Dian Fosse, and Cynthia Moss all started their field research in the 1960s and 1970s. Technology for wildlife barely existed during that time. The best and often only ways to study animals were by directly doing the fieldwork yourself: watching animals, trapping animals, or surveying them indirectly by counting scat/dung, burrows, and/or listening for vocalizations. The data you collected for a single project was typically a sample size in the dozens or hundreds. Data entry was written down into field notebooks and done by hand.

Even when technology did emerge for wildlife applications, such as GPS trackers and camera traps, the sample size was still limited. For GPS telemetry, researchers still had to go into the field to get fixes, and film camera traps would only take up to 36 photos when deployed in the field.

Given the smaller sample sizes and limited computing power at the time, the statistical tests scientists performed were manageable. When I was in graduate school, one of the older professors in our department made his first figures by hand. It's completely different today.

Technology has changed wildlife biology in two major ways. First, the use of sensors like camera traps, GPS tracking devices, and acoustic recording devices deployed in the field means that scientists no longer have to rely on humans as much to collect data. Sensors have a lot of advantages over humans: they work day and night continuously, collect orders of magnitude more data, and allow scientists to study animals without human presence, reducing biases animals may have towards people.

Using sensors frees up time for scientists allowing them to do more difficult tasks that cannot be automated, like analyze data and write manuscripts. While researchers still have to go to the field to deploy sensors, battery life has improved a lot, requiring fewer field trips. There are even solar options requiring no batteries.

As a result of these improved technologies, wildlife biologists are collecting a lot more data. These sensors, in addition to other advancements in technology, such as genetics and genomics research, mean that scientists are no longer analyzing data on the scale of hundreds of data points but rather thousands, millions, and even billions when conducting meta-analyses. A single camera trap deployed in the field can last for months and collect tens of thousands of photos. Sophisticated GPS technology can collect data on individuals every minute or even seconds apart. Scientists are no longer even looking at single genes but conducting genome-wide studies as it gets cheaper and easier to sequence such data.

To make things more complex, these are only individual samples in themselves. One camera trap, one tracking device on an animal, and one genome all represent one animal individual. To conduct research, you need multiple samples. In the camera trap research that I participate in, 20 cameras traps are the bare minimum we will use to conduct a proper study.

Additionally, scientists are turning towards citizen science and crowdsourcing to collect data on massive scales and in massive amounts. The majority of people worldwide have a smartphone with built-in cameras, even in rural areas of developing countries. These phones make it possible for citizens to capture photos of their local environment, make observations, and upload data in seconds. eBird, a popular citizen science program for birders, has half a billion data points and hundreds of thousands of users across the world as I write this.

Science is also trending towards open access databases where researchers deposit data about the species they study. These databases enable scientists to conduct meta-analyses with much larger sample sizes. Movebank, a database that houses the location points of tracked animals, contains over a billion location points. eMammal, the citizen science camera trapping project I work on, contains millions. eMammal data now feeds into Wildlife Insights too. The goal of Wildlife Insights is to unite camera trap data from around the world into one central database. Seven of the world's leading conservation and science organizations have already agreed to include their data sets.

The quote from this scientific paper published in 2020 by Isaac *et al.* explained this perfectly:

*"The information available for models of species' distributions is radically changing, thanks to a digital and technical revolution in data collection. New technologies, such as camera traps, miniature geolocation devices, environmental DNA (eDNA), and passive acoustic monitoring, are creating new opportunities for surveying wildlife in space and time. These developments, allied with initiatives for data mobilization and the rapid growth of citizen science, mean that ecological data are being generated at an unprecedented rate, in an ever-increasing number of formats and currencies."*[3]

Technology is also reducing the number of steps scientists have to take to extract numbers from raw data. This in turn reduces employment opportunities for field technicians and entry-level positions. Remember how in graduate school, I had to sort through thousands of forest elephant photos by hand to identify individuals? This process of identifying individuals has largely been replaced or sped up by technology.

Computer scientists are collaborating with wildlife biologists to create programs for identifying individual zebras, giraffes, spotted cats, and whale sharks, to name a few. All you have to do is upload a photo and the program gives you back the individuals that match the photo best and how confident

---

[3] Isaac *et al.* 2020. Data integration for large-scale models of species distributions. *Trends in Ecology & Evolution.* 35: 56-67.

the program is in the identification.

There is now technology being developed to identify even moving animals from drone footage or video camera traps to categorize the postures of individuals in behavioral studies. At Disney, I remember the animal behavior intern talking about how her main job was to watch hundreds of hours of animal videos and take note of the different behaviors the individuals were doing. Now engineers are developing software to automate this and save scientists hundreds of hours of work.

At eMammal, we currently manually verify each species identification from photos uploaded by a volunteer or researcher. With some projects having thousands of camera trap locations, this can take researchers a long time. To solve this problem, in Wildlife Insights, we collaborated with Google to develop artificial intelligence (AI) to automatically identify species. While AI will never replace scientists entirely for species identifications as some species are difficult to distinguish between even for scientists, it will dramatically reduce our workload. In North Carolina where I work, most photos consist of white-tailed deer. The AI performs well on white-tailed deer and when the species choices are restricted to the state, it cuts our workload by more than half. On average, white-tailed deer make up at least 50% of our animal detections.

This is very good news for conservation because it greatly shortens the time for scientists to obtain data, meaning that they can understand species' abundance and distribution patterns in near real-time. Camera trap researchers are often so backlogged with data that it takes months or even years before data analysis can begin. If a scientist's research question is about the efficacy of a protected area, years might be too late. The results may be obsolete when they are released if, for instance, there is heavy poaching in the park.

With the increase in data and push for the availability of data through open access sharing platforms, much more emphasis has been placed on data analysis in wildlife biology. Statistics has always been an important part of science, but before you could be successful with knowing the basics as computing power was limited and you even had to do the work by hand before computers were common. Now everyone in graduate school owns a computer capable of doing powerful analyses, and there are labs available on campuses dedicated to handling massive amounts of computing power.

As I mentioned in the first chapter, I have seen dramatic changes take place even in my experience since beginning graduate school in 2006 to most recently as a postdoc. Then, students were using SAS® software for statistics, or just starting to conduct analyses in R®. Now, R® is the go-to program that everyone uses. Since then, hundreds if not thousands of packages have been developed for advanced statistical modeling in R®.

Nowadays, students are not only running existing code for their research projects but creating their own models, pushing the boundaries for what can be done. One of my lab mate's statistical modeling is so complex and computationally demanding that she uses the "super computers" at NC State University to run them. These computers allow her to run her analyses for days at a time. In other words, a single complex model may take days or even over a week for the computer to process it and report the results.

*Camera traps like this one can now hold tens and thousands of images on a single memory card.*

In order to be a wildlife biologist, loving animals and nature are not enough. You really need to love, or at least like statistics. I remember attending one talk at the annual Wildlife Society Conference where a professor presented a new advanced statistical model. The talk was so packed, there was standing room only.

In response to a tweet I wrote on the importance of statistics in wildlife biology, someone answered that a researcher in their lab studied moose using camera traps. They got their master's degree without ever seeing a moose in person. This was actually the case with my Ph.D. advisor too. She studied forest elephants non-invasively by using their dung and never saw one during her fieldwork. She saw her first forest elephant when scouting for field sites with me.

Through my experience looking on the job boards for the past few years, many postdoc positions desire advanced modeling skills. Because the field has changed so much in recent years, when asking people for advice, I think it is best to ask people who have recently acquired permanent jobs for their perspective. Professors and professionals who obtained their degrees more than a decade ago will likely not understand what is needed for jobs today in this competitive environment.

# 6 OVERVIEW OF CAREER TYPES

Now that we know what a wildlife biologist is, let's go over more specifically what they do, and other career types related to the field of wildlife biology. You may think it's strange for me to jump into discussions about jobs when we haven't addressed education yet. The reason why I am working backwards is because if you don't do the proper research for what job is right for you, you might end up spending time and money on courses or experiences you don't need, or even worse, getting the wrong kind of experience or becoming overqualified for the job you want.

There are a lot of positions that are not research-based but still in the realm of wildlife biology and incredibly fulfilling as a vocation. It's likely that you are unsure of your career direction like I was. Getting a bird's eye view of the options available may help you eliminate some categories you are not interested in and perhaps narrow your list. The important thing to keep in mind is that if you think you could possibly want a specific career type or to work in a specific workplace, you need to make sure you get the skills necessary to be able to pursue that job should you actually do so. It is better to be overprepared (but not overqualified) than underprepared.

At this point, I also encourage you to start looking at the job boards and use the free job tracker spreadsheet that I created. In the course notes, I link to a blog post I wrote that summarizes and connects you to the major job boards in this field. The job tracker tool, available for free at https://stephanieschuttler.com/job-tracker/, will help you pay attention to what types of jobs you are interested in, see how often they are available, and what skills and experience you will need to get these types of positions. You may surprise yourself. Going into this, you may think you want to study grey wolves for the government, but instead realize you would be much happier talking about that research to the public.

This chapter will provide a very broad overview of those career types.

There will be additional types also tangential to the field that I will briefly discuss at the end, but in my opinion, these are the main ones. For a more exhaustive list of job position titles and places of work, especially in the government, I suggest reading the book: Becoming a Wildlife Professional, edited by Scott E. Henke and Paul R. Krausman. A link to purchase it is in the course notes.

For each career type, I provide possible titles (not exhaustive), likely responsibilities, the types of workplaces that hire them, the degree(s) you will need, and questions you should ask yourself to help you determine if this career is right for you. If you answer yes to most of them, that is a good indication that this might be a career option for you.

Please do not limit all of your knowledge of a career based on these descriptions. I encourage you to seek out other sources for information, especially from people hired in those careers (best if they are recently hired). This is meant to give you a rough guide to provide you with some direction as to where you should go or what is available to you.

Please note that my experiences and knowledge are from working with institutions based in the United States, even if interacted with them internationally (e.g. WCS in Gabon). Therefore, this list is centric to jobs in the US. Other countries will have their own federal agencies and potentially state/province agencies, of which case you will need to research the qualifications necessary to obtain positions within them. Different countries may also have their own requirements for securing jobs. For instance, you may need a country-specific certification.

Of the career types listed below, I have no direct experience working in the law or policy category. Therefore, my knowledge is limited in this area and I suggest you should seek out resources that specialize in this field.

## Researcher

### *Possible Titles*:
Wildlife biologist, research scientist, mammologist, furbearer biologist, ornithologist, herpetologist, park or refuge manager, ecologist, wildlife ecologist, research technician, field technician, research specialist, field specialist. This list is not exhaustive. Note: Technician positions are often temporary.

### *Responsibilities*:
The primary focus of wildlife biologists is to ask applied and/or theoretical questions about wildlife, collect data to answer those questions, analyze the data, and write up the results in technical reports and/or peer-reviewed publications. What you do daily will vary according to how much education you have and where you work.

Wildlife biologists with bachelor's degrees are more likely to spend time in the field doing data collection and entry, especially when starting out, while those with advanced degrees will be doing more data analysis and writing, as well as attending meetings within the organization and other stakeholders.

At the other extreme, if a wildlife biologist has a Ph.D., they almost always have more of a management role supervising field technicians or other scientists in data collection. They are less likely to work in the field and are more likely to manage several research projects or even oversee all of the research within an organization's office or region (for example, the science director for the Nature Conservancy in a state). At the Ph.D. level, you will be driving the research and likely have your own lab. Therefore, you will be deciding specific research questions (within the boundaries of your organization's mission), designing projects, and writing grants to fund them.

Positions for wildlife biologists with master's degree will typically fall somewhere in between. They will likely have some field work, but also manage field technicians/staff, as well as indoor office work such as analyzing data and writing reports. They may have some role in designing research, writing grants, and/or be able to choose their own projects, but they are likely to be overseen by a more senior scientist.

The nature of a wildlife biologist's research questions will vary according to where they are employed. We go into detail on this in the next chapter. If you work at a government agency, consulting agency, zoo, nonprofit organization, or for tribal nations, your research will be very applied, such as conducting surveys for wildlife management and conservation. If you work at a museum or as a professor, your work will likely be more theoretical.

*Where Do They Work:*
Academia (universities and colleges), government (state and federal), tribal nations, nonprofits, zoos and aquariums, museums, consulting agencies

*Degrees Needed:*
Bachelor's, but most jobs now require a master's or Ph.D.

*Questions to Ask Yourself:*
The work that you do day-to-day will vary a lot according to the degree that you have and where you work. Those with higher degrees will be indoors more, conducting more analytical research, while those with more entry-level positions will have more fieldwork and data entry. However, the following questions should give you some indication if you enjoy research.

Do you like:
- Asking questions about the natural world? Do you find yourself wondering why something is the way it is? Are you innately curious?
- Solving problems?

- Managing spreadsheets full of numbers? Are you okay with repetitive tasks? Tedious tasks?
- Imagine you were given a spreadsheet full of data. Would you find joy figuring out how to analyze the data and actually conducting the analysis?
- Using technical computer programs?
- Reading academic papers? Give it a try. I have a YouTube video on how you can find them for free. Can you imagine yourself reading papers like these on a near daily basis? A lot of science is reading and writing. Would this bore you to death or interest you?
- Do you like writing reports or scientific papers? A large portion of research is writing.
- Do you like working alone? While there is definitely teamwork involved in research, a lot of the writing and analysis is done alone.
- If you had a million-dollar grant, what kind of research project would you do?

The people that I know that love research, absolutely have dream projects. In non-work hours, they are thinking about other questions they could ask and how they could make their research better. I, on the other hand, was largely not driven by research. I was driven more by solving conservation problems. I didn't love coming up with questions; I would much rather have an organization tell me that they needed help with a specific species or protected area so I could develop applied questions around those subjects.

Additionally, what type of project you come up with can also give you insight on what kind of research you like and therefore what types of workplaces might be good for you. For example, if you come up with a conservation project, you might be more suited for nonprofit and government research than academia.

**Professor and Lecturer**

*Possible Titles*:
Assistant professor (title for new professors), lecturer

*Responsibilities*:
Professors teach courses to undergraduate and/or graduate students, conduct research, and mentor students doing research. Tenure track positions are those where once you are awarded tenure, your job is secure. Non-tenure track positions typically have more of an emphasis on teaching. In lecturer positions, you only teach and do not have your own research labs.

For professor positions, your time will be split between research and teaching. This will vary according to the type of university that you work for, R1 or R2.

R1 (research 1) universities place more emphasis on research. In these positions, you will be expected to publish peer-reviewed papers regularly and acquire large grants (hundreds of thousands or even millions of dollars). You will still have to teach and be expected to perform well at teaching (i.e., get good evaluations from students), but whether you get tenure or not is primarily evaluated by your research output (e.g. publications and grants).

At R1 universities, you will be expected to publish at a faster rate and generally in higher impact journals. The exact publication rate expectations will vary according to your university, but you should expect between two to four peer-reviewed publications a year and to teach one to two classes.

As a professor conducting research, your research experience is different from how you experienced research as a graduate student. Instead of going to the field, collecting data, and processing your samples, your students will be the ones who do those steps. It is now your job to mentor students through this process from project start to finish. The goal is to not only have students complete their thesis or dissertation but also to make sure they publish their findings in peer-reviewed journals. Publications from your students counts as one of your own as you are the last author on this paper, signifying that it is from your lab. This is not to say you won't be conducting your own studies as a professor, but a lot of your publications will be from your students' research. You will also serve on other students' committees and help mentor them.

Professors at R1 universities also have the opportunity (or expectations) to mentor postdocs. Postdocs should take less time to mentor because they already went through the Ph.D. process, should have a publication history, and are therefore better adept at hitting the ground running. However, as a professor mentoring a postdoc, you are more responsible for helping them prepare for and secure a permanent position afterward compared to a graduate student.

Professors at R2 (research 2) universities conduct research, but more emphasis is placed on teaching than at R1 universities. Professors at R2 universities teach more classes and tenure is more heavily influenced by teaching performance. The research and publication expectations for R2 universities are usually lower because of the higher teaching requirements. Again, this will vary across universities, but roughly, expect to write two peer-reviewed publications while teaching two classes a year at R2 universities. R2 universities tend to have only undergraduate or master's degree programs. Keep in mind that mentoring undergraduate students will require more one-on-one time than graduate students because graduate students are expected to have more independence, self-motivation, and time for their research.

Research at R1 and R2 universities can be applied and/or theoretical. Professors need to secure funding to conduct their research which requires grants. New professors are provided with some startup funds. Large grants such as those awarded by the National Science Foundation require theoretical research. However, you can sometimes spin applied research within theoretical questions. For applied research, professors can secure funding through private donors, the state wildlife agency, and smaller grants.

Professors may teach classes as small as ten or fewer students or as large as hundreds of students. They may teach at both the undergraduate and/or graduate level. Teaching consists of creating a course syllabus, writing lectures, assigning readings, developing tests and/or writing assignments, and grading materials. Courses with lab components will likely have graduate teaching assistants to teach the laboratory portion, but as the lead professor, you will have to manage the teaching assistants.

Teaching is a lot of work. When I was a lecturer for mammalogy, I was surprised at how long it took me to create a single lecture from scratch, approximately eight to ten hours. My colleagues who are now professors tell me it is much easier if you teach the same course repeatedly, which is what often happens.

### *Where Do They Work*:
Universities, colleges, and community colleges. If you love teaching, you can also get good positions teaching high school science, especially in private schools.

### *Degrees Needed*:
Master's or Ph.D. A master's degree can be acceptable for a community college or as a lecturer, but all other positions should require a Ph.D.

### *Questions to Ask Yourself*:
Do you like:
- Teaching? Try a semester as a teaching assistant. Do you like designing courses, making up lectures, writing and grading exams? Coming up with innovative ways to engage students and reach students who learn in different ways?
- Mentoring students? Helping them with the next steps of their research? Advising them and aiding them when problem arise?
- Reading and writing scientific papers?
- Managing students? If you are at a R1 or R2 institution you will have a research lab and have to manage undergraduate and/or graduate students, how they use the lab, and how they work together.
- Being in an academic setting? Regularly attending seminars and

academic discussions?

## Educator

***Possible Titles*:**
Education officer, environmental educator, outreach project officer, team/field leader, education instructor, interpreter, naturalist, outreach coordinator

***Responsibilities*:**
An educator's job in the realm of wildlife biology is to convey messages of wildlife obviously, but also science and conservation to the public. The public will consist of multiple audiences depending on what organization you work for, what your specific job title is, and/or the nature of the event you are working on at the time. Often the focus of outreach is related to the research conducted at the institution you are working at.

Educators' responsibilities may include activities such as guiding nature walks, overseeing activities that involve wildlife (e.g., birdwatching, using nets in ponds to find animals), writing lesson plans for schoolteachers, visiting classrooms, and speaking at events. There are also many jobs in education where you spend a lot of time outside.

An educator is a great job for people who love to communicate and be around other people. I've enjoyed my education and outreach experiences both at Animal Kingdom and NCMNS. One of my favorite things to do was talk to the public about the wildlife research we were conducting.

Educators in higher-level positions lead and manage staff. Much of their time is spent in meetings planning and executing the vision of the institution where they work.

***Where Do They Work*:**
Government (state and federal), tribal nations, nonprofits, zoos, and aquariums, museums. There may be jobs in academia. For instance, NC State University runs a division called the Science House which creates STEM (science, technology, engineering, and math) programming for K-12 teachers.

***Degrees Needed*:**
Bachelor's and maybe a master's degree. For many, but not all, jobs, you will be overqualified with a Ph.D. If you want an education position, it may even be better for you to get a degree in education instead of or in addition to a science degree, especially in environmental education.

***Questions to Ask Yourself*:**
Do you like:

- Conversing with the public? Giving talks and interacting with people one on one?
- Making educational materials? Writing blog posts? Designing brochures, fliers? Coming up with creative ways to engage or teach people?
- Thinking of games or activities to convey educational messages?
- Organizing and leading groups? A big part of education is often logistics.
- Planning and setting up events? You may be putting on your own events or tabling at a festival.
- Being around other people? Answering questions? Even the same questions over and over again?
- Are you high energy? Can you maintain enthusiasm and a positive attitude no matter what?

## Communications Specialist

*Possible Titles*:
Social media manager, communication manager, science communicator, communication specialist

*Responsibilities*:
Most organizations now have communication specialists given the heightened importance of social media in the last two decades. Communication specialists convey the mission of the organization they are working for to the public by highlighting essential projects, people, and research done by the organization. They run social media accounts, update websites, write blogs, and advertise events.

Just because you are savvy on social media and have a big following does not necessarily mean you are competitive or qualified for such positions. From what I've seen of these types of position advertisements and applying to some myself, organizations are more interested in hiring communications professionals who are interested in science rather than scientists who do science communication in their spare time. Frequently these jobs are advertised with communications or marketing degrees as a requirement.

To be competitive for these types of jobs, you will likely need to know how to develop a social media plan, have a process for evaluating its effectiveness, and for nonprofits, know how to develop and run social media fundraising campaigns. Communications jobs require you to be literate in things like Google Analytics, marketing, advertising, and each social media platform's analytics, to name a few.

There are also jobs in communications where you are the liaison between the press and the organization in which you work for. In these jobs, your responsibilities include writing press releases and coordinating stories with the media. A communications or journalism degree is more appropriate for these types of positions.

### *Where Do They Work*:
Academia, government (state and federal), tribal nations, nonprofits, zoos and aquariums, museums, consulting agencies.

### *Degrees Needed*:
Bachelor's and maybe a master's degree. A Ph.D. will likely over qualify you for these kinds of positions. You may consider getting a marketing or communications degree instead of a science degree for these types of jobs.

### *Questions to Ask Yourself*:
Do you like:

- Posting to social media? Coming up with creative ways to engage the public digitally?
- Developing social media strategies? Analyzing social media traffic? Finding patterns to make more successful campaigns?
- Writing blog posts or designing websites?
- Speaking and/or organizing with the press?
- Responding to social media messages? Both public and private?
- Running campaigns? Fundraising efforts?
- Synthesizing and organizing information? You will likely have to reach out to different researchers and/or divisions within your organization to gather relevant and interesting updates.

## Writer/Filmmaker/Photographer

### *Responsibilities*:
Writers, filmmakers, and photographers tell stories about wildlife, research, and/or the scientists that study wildlife through their media. By story, I do not mean fiction or fantasy; I mean organizing factual information and presenting it through a compelling narrative. The BBC series *Planet Earth* does an excellent job of this, taking you through the lives of glow worms living in caves to savanna elephants in the Namibian desert, while incorporating important scientific topics like evolution.

Writers, filmmakers, and photographers all obviously work with different media. They will, therefore, have different technical skills. There is a lot of variation in the types of jobs even within the same medium. For example, within in writing, science journalists may write stories without opinions on topics and recent discoveries in science (e.g., Carl Zimmer for the *New York Times*), authors may write books (again Carl Zimmer), and/or write popular pieces for a public audience such as a magazine or blog (e.g., Ed Yong for *Scientific American*). If you want a career as a writer, most jobs require you to extend your writing beyond wildlife biology and cover science as a whole.

Photographers and filmmakers take professional photos/videos of wildlife and/or make documentaries and shows about wildlife for television, film, and/or the Internet (e.g., YouTube). In addition to being a good storyteller, you will need technical skills for photography and video.

Permanent positions in all of these fields are becoming more and more scarce as newspapers and magazines have downsized over the years. Many content creators make their living through freelance work where they will create specific products for companies through independent contracts. For example, if a nonprofit wanted a video made on a research project, they would contract this work out to a filmmaker. In addition to creative skills, you will also need marketing skills to sell yourself and your work as a freelancer.

***Where Do They Work***:
Newspapers, magazines (e.g., *Discover, National Geographic*), science websites (e.g., *Scientific American*), television (e.g., *Discovery*), and YouTube channels. Most that I know are freelance and contract their work out.

***Degrees Needed***:
Technically none, but a bachelor's degree will perhaps serve you well. For a career in writing, it is probably better to major in English or journalism, and for a career in film or photography, a major or minor in those media. To pursue these careers, you do not need a degree in wildlife biology, but having a solid understanding will help you create better content and products.

***Questions to Ask Yourself***:
Do you like:
- Being creative?
- Finding the story in big picture problems or in research?
- Constructing a narrative?
- Selling yourself and your products?
- Writing?
- Do you have a specific point of view?
- Using technical equipment (for film and photography)?
- Working for yourself?
- Would you be comfortable with the idea of not getting a steady

paycheck? Are you good at organizing your finances so you have money in times where you may not be able to secure work?

**Law and Policy:**

*Possible Titles*:
Policy advisor, policy analyst, wildlife enforcement officer, special agent

*Responsibilities*: After drugs and guns, wildlife is the third most illegally trafficked item in the world. This business runs in the billions of dollars. Think poaching only happens in Africa with elephants and rhinos? It doesn't; it can happen anywhere.

Wildlife agents are needed to enforce the laws already in place. These agents frequently work on the ground, for instance in state and national parks, to ensure the public abides by the rules. Some of their responsibilities are even similar to police officers and do not involve wildlife, for example, handling park-goers when they are excessively noisy or causing disturbances at campgrounds.

Conservation and government organizations also hire people to lobby for stronger wildlife protections and for consulting on regulations involving wildlife. To be competitive for these positions, you should have a strong understanding of national and/or international wildlife laws. You will also need a working knowledge of wildlife biology as you will have to understand how different policies will affect wildlife populations. For example, there is a lot of debate over whether legalizing rhino horns and elephant ivory will decrease or increase the rate of poaching. It is necessary to understand the principles of wildlife biology to be able to work alongside scientists in projecting how such policy changes could impact vulnerable species.

Another field related to wildlife law is wildlife forensics, where scientists help solve crimes involving wildlife. These positions usually require training and experience in genetic research. Often the work involves trying to determine where confiscated wildlife goods originate from to know where protections need to be strengthened to reduce poaching. Wildlife forensics is a pretty niche field, meaning there aren't many positions, but it is an interesting career option.

*Where Do They Work*:
Government (state and federal), tribal nations, nonprofits.

*Degrees Needed*:
For law enforcement, a bachelor's degree. For policy and wildlife forensic jobs, at least a master's degree, but probably a Ph.D. for some positions. Some jobs may require you to have courses or degrees in policy. Some

positions may even favor those with a law degree.

## *Questions to Ask Yourself:*
These questions will vary a lot according to what type of position you have (e.g. law enforcement or policy). Do you like:
- Understanding rules and regulations?
- Working with a variety of stakeholders?
- Being in long meetings and representing an organization?
- Negotiating or making compromises?
- Public speaking?
- Paying attention to small details?
- Are you patient? Policy changes can take years or even decades of work.

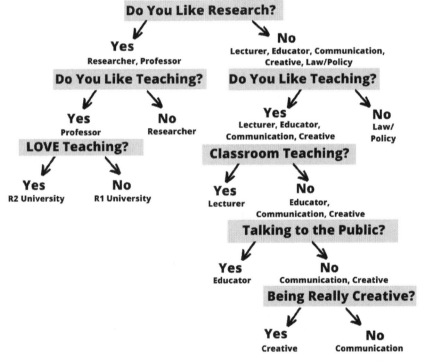

*A flow chart I created to help you determine which career type is best for you. Creative refers to writer/filmmaker/photographer jobs.*

As mentioned before, this is a general overview of the categories of jobs related to wildlife biology. It is not exhaustive, and positions evolve over time (e.g., careers in social media). Additionally, if you really enjoy working with a particular program such as ArcGIS® or technology such as drones, you may be able to find a job dedicated to those technologies.

Some titles are more difficult to place under a specific category and may fall under multiple categories. For example, coordinators or land managers could be placed in the categories of research, education, and policy, even within a single position. A land or park manager may require a research background to interpret data and make decisions about the wildlife on the land, public speaking and effective science communication skills to hold meetings with the community, and policy skills in working with the local government.

Another career involving wild animals that I will not go into depth with, but you may want to look into, is those in the ecotourism and wildlife tourism sector. Some of my friends have careers as guides leading trips all over the world, and some even run their own ecotourism operations.

I did not include wildlife rehabilitation experts, sanctuary employees, and zookeepers, even though those career types also work with wildlife. These may be excellent career options for you, but I do not cover them in this book because these careers do not fall under the context of wildlife biology. Their focus is on animal husbandry and captive animal management rather than research on wild animal populations. In those fields, more emphasis is placed on care and veterinary medicine as the health and wellbeing of individual animals is the primary concern.

Now that we have a rough overview of the job categories within the realm of wildlife biology, let's explore the types of workplaces where you can get these jobs.

# 7 CAREER WORKPLACES

Wildlife biology and its related careers can take place at all sorts of institutions that vary in their missions. Some organizations emphasize education and outreach, while others are strictly research-based. This is a coarse overview of the types of places where you can get a job in wildlife biology and what types of jobs you may find there. Knowing the kinds of organizations you want to work for will better prepare you for that job when you are ready to apply. Remember from Chapter 3 that you can't pack for a trip if you don't know where you are going. Different workplaces have different cultures, require different qualifications, and experiences. Of the places listed below, I do not have direct experience working in tribal nations and consulting agencies, and therefore my knowledge is more limited in those work environments.

## Academia: Universities and Colleges

Universities and colleges are institutions for higher learning. They vary dramatically in size from hundreds to tens of thousands of students and offer a variety of courses and programs. Some schools have specific wildlife programs and departments, while in others, courses may be embedded in a more general department (e.g., biological sciences).

Colleges and universities vary in the degrees they offer. Some may only have undergraduate programs, or undergraduate and master's programs, while others have all graduate degree options.

By far, most career options in academia are professor or lecturer positions, but there are also research positions within labs and extension biologists. Extension biologists work for the university, but they also work with the federal and/or state government. These jobs are more research-focused with few, if any, teaching responsibilities, and are permanent positions.

Research positions within labs (e.g., lab manager, research associate, technician) are temporary because the salary is usually generated from specific grants. Professors that conduct research in a lab (e.g., genetics) though may have a full-time lab manager.

In recent years, universities' science or wildlife departments have started to hire communication specialists to develop social media and other forms of outreach. There may also be positions available in public education at universities. For example, NC State University has an educational extension program, the Science House, focused on STEM outreach to K-12 teachers.

*Examples of Job Types:*
Professor, lecturer, advisor (to undergraduate students), extension biologist, positions in science communication and education

*Where:*
Any college or university. There are too many to list across the country!

## US Government: State and Federal

This section is written specifically for jobs in the US government; however, the information may apply to agencies around the world. You will have to do your research; it is impossible to summarize information for every government, but you can use this information as a starting point.

Two levels of government manage wildlife in the United States: the federal and state governments. The federal government operates at a national level, while each state has its own fisheries and wildlife agency (they will vary by name) that is responsible for conserving and managing wildlife for the public good. State governments actually play the largest role in managing wildlife.

The federal government has many agencies that hire wildlife biologists (these are listed below, under "where"). The agency most applicable to wildlife careers is the US Fish and Wildlife Service, whose purpose is "to conserve, protect and enhance fish, wildlife, and plants and their habitats for the continuing benefit of the American people." Their work differs from wildlife management at the state level in that they usually work with federally listed endangered species and/or species whose range involves many states, for instance, migratory birds.

State agencies are responsible for the overall management of all wildlife within their state. Historically, state agencies focused exclusively on game species but are increasingly placing attention on non-game wildlife, with many agencies dedicating entire positions devoted to non-game species. If you are a wildlife biologist at an agency, determining hunting and trapping limits will likely be one of your primary responsibilities and a large focus of the job. To work at a state agency, you must be comfortable with promoting hunting and trapping even if this is not part of your job. This is the primary way state agencies make money, and hunters, trappers, and fishers make up most of the agency's actively participating constituents.

If you are interested in working for the government, it pays to start working with them as soon as possible. Both state and government jobs have a formal application that can be difficult to qualify for, even if you are an excellent candidate because of bureaucratic technicalities (see the conversations from Twitter in the course notes for examples). For the federal government, once you are in their system, it is a lot easier to get rehired for other positions.

*Examples of Job Types:*
Wildlife biologist usually, but there may be many different job titles that fall under wildlife biologist. Refer to the research positions section for a more extensive list of job titles. Federal and state governments also hire people to work in law enforcement, science communication, education, and outreach.

*Where:*
Federal government: US Fish and Wildlife Service (USFWS), National Park Service (NPS), US Geological Survey (USGS), Bureau of Land Management (BLM), US Forest Service (USFS), the US Department of Agriculture (USDA), US Army Corps of Engineers, National Oceanic and Atmospheric Association (NOAA), and US Bureau of Indian Affairs. These are the most common divisions to have wildlife biology jobs, but there may be more in other agencies. A more exhaustive list as well as a description of each agency is located in *Becoming a Wildlife Professional* by Henke and Krausman.

For the state government, each state will have their own agency and they will vary by name. For example, the North Carolina Wildlife Resources Commission is the state agency for North Carolina, while the Department of Environmental Conservation is the state agency of New York.

**Tribal Nations**

In the United States, American Indian tribes act as sovereign nations. Therefore, they have the authority to manage their land and the wildlife within their reservations independently. American Indian reservations make up over 200,000 km$^2$ of land in the US. Like the state and federal governments in the US, they hire wildlife biologists to survey, manage, and conserve their wildlife. The role of the wildlife biologist is to manage wildlife for the tribe rather than for the public.

You don't need to be a member of the tribe that you are working with or an American Indian to be a wildlife biologist for a tribal nation. Still, as part of your job, you will be integrating the tribe's traditions, culture, and attitudes towards wildlife into the management plan in addition to scientific advice. Like a wildlife biologist for the government, your work will likely focus on game species and include responsibilities like setting harvest limits and hunting seasons. These may differ significantly from those set by the government and even what you learned in school because it is necessary to incorporate the tribe's culture.

Sometimes laws set by the US government come into conflict with the culture of tribes. When this happens, decisions are made in court. For instance, some tribes use eagle feathers for cultural traditions and dress. However, there is a US law prohibiting the take and ownership of such feathers because species have historically suffered from significant declines.

The take of eagle feathers by tribes though is and has always been incredibly small. It's also something that tribes partook in for centuries without affecting eagle populations. Therefore, the courts ruled in the tribes' favor to continue this tradition. As a wildlife biologist working for a tribal nation, your job is to integrate the tribe's traditions into wildlife management protecting both the tribe's culture and wildlife.

*Examples of Job Types:*
Wildlife biologist or similar, potentially positions in education.

*Where:*
American Indian Reservations

## Nonprofits or Non-Governmental Organizations (NGOs)

Nonprofits or non-governmental organizations (NGOs), hereafter referred to collectively as nonprofits, are organizations founded for a specific cause and are, therefore, by nature, advocacy driven. The missions of nonprofits in wildlife biology are typically focused on the conservation of wildlife through scientific research, education, and/or public outreach.

For nonprofits to carry out their mission, they have to continually fundraise by securing large individual donations, running public fundraising campaigns, and/or writing grants. Many people who work for nonprofits often have to fund their own salaries, usually through grants.

Nonprofits hire wildlife biologists to research the conservation and management of all different types of wildlife and their habitats. This frequently involves program management and coordination, engagement with various stakeholders, and conservation planning.

In my experience looking for jobs at nonprofits, there are fewer research positions and many more jobs available in fundraising, communications, marketing, and development. Larger organizations like the Wildlife Conservation Society, World Wildlife Fund, and Conservation International often focus their conservation efforts in biodiversity hotspots, which are frequently located in developing countries. Most of the research positions will therefore be located in these countries so the scientists can conduct in-country work more easily. Most of the jobs based in the US for those organizations tend to be those in fundraising, marketing, etc.

If you are interested in working for a nonprofit, I suggest you acquire skills such as outreach, community-based conservation, and fundraising, which are important to nonprofits to make you a stronger candidate. Remember that when I interviewed for an executive director position where I thought I would have been a perfect fit? I didn't get the job because I didn't have any fundraising experience. If I had volunteered on a fundraising campaign at an organization or taken fundraising courses during graduate school, I may have been chosen for the position.

Nonprofit careers tend to be less stable because of the volatility in fundraising. During tough times, they may cut jobs. As I mentioned before, there are also many positions where you have to secure your own salary through grants or other funding sources (called soft money). Some people find this to be very stressful, eventually get burned out, and leave.

*Examples of Job Types*:
Wildlife biologist, positions in education and outreach, science communication, policy

*Where:*
World Wildlife Fund, Wildlife Conservation Society, Conservation International, the Nature Conservancy, the National Wildlife Federation, and the Audubon Society are examples of some large international and national nonprofits. There are many smaller regional and ones, too many to list here.

## Zoos and Aquariums

Zoos and aquariums are institutions that have collections of live animals on display for the public. Decades ago, these collections were originally created for entertainment, but zoos and aquariums evolved over time and now play a large role in public education and conservation of wild animals. They are especially important in reintroduction programs. When species or populations go extinct in the wild, provided there is enough habitat in place, reintroduction programs can reestablish populations.

Captive breeding and the housing of animals make up a large component of zoos and aquariums' mission. There will likely be ongoing research on reproductive biology, animal behavior, endocrinology, genetics/genomics, and disease ecology on the animals at the zoo or aquarium. This type of research may require a stronger background in zoology than wildlife biology.

The animals at zoos and aquariums serve as educational "ambassadors" for the species in the wild, as most people will never see them there (including wildlife biologists like myself). Many institutions not only research and care for animals at their facility, but also their wild counterparts. There is a wide range of research programs within zoos and aquariums, with some having minimal or no research funds, while others having extensive, world-class research programs.

Zoos and aquariums, by definition, engage with the public. If you want a job at a zoo or an aquarium, it's an excellent idea to get some skills/training in science communication even if you desire a research position. In my opinion, it's best to work at institutions that have accreditation by the Association of Zoos and Aquariums (AZA). The AZA is a nonprofit organization "dedicated to the advancement of zoos and aquariums in the areas of conservation, education, science, and recreation." AZA institutions usually have higher levels of animal care and standards for welfare.

*Examples of Job Types*:
Wildlife biologist, positions in communication and education

*Where*:
All over! Too many to list. Some that are known for their research: San Diego Zoo, Bronx Zoo (run by the Wildlife Conservation Society).

## Museums

When you visit a natural history museum, for displays on wildlife, you will usually see skeletons and taxidermy on exhibit for people to learn about the animals and their habitats. However, museums have a deeper purpose that is hidden from the public eye. Within every museum are large collections of specimens from fossils millions of years old to the skeletons and skins of different taxa today. These specimens have been collected for hundreds of years. They provide a snapshot of what the species was like at a specific point in time.

All museums have curator positions to oversee these collections with their own research programs. Curator research involves specimens, so usually, their research is focused on taxonomy and understanding the evolutionary history of species. Because of the ability to obtain DNA from museum specimens (e.g., hair, feathers, tissue samples), often curators today are focused on genomic research, whereas historically, most research was based off of morphological measurements of skulls and skeletons.

If you work directly in the museum collections (either as a curator or collections manager), some of your responsibilities may include overseeing the loaning of specimens to other researchers, preparing specimens (for mammals and birds, skinning and stuffing them), and going on collection trips (trapping and sacrificing animals to expand the museum's collection). Some museums (like NCMNS) also hire scientists to do research outside of specimens.

Education is a vital component of most museums' mission because they are usually open to the public through the exhibits that you see. If you desire a job at a museum, it is therefore, a good idea to get some skills/training in science communication. Many museums now have regular events where scientists speak to the public.

*Examples of Job Types*:
Wildlife biologist, curator, collections manager, positions in communications and education

*Where*:
All over! Too many to list. Some that are known for their research: American Museum of Natural History, the Smithsonian's National Museum of Natural History, California Academy of Sciences, Chicago's Field Museum.

## Consulting Agencies

Consulting agencies are private companies that employ wildlife biologists where they are then contracted out to do specific projects. Other organizations, such as the government, nonprofits, industries, and developers, will hire consulting agencies to do specific, temporary projects.

Usually the company that is contracting out the work wants to modify or develop the landscape. They will employ the agency to do the work necessary to make sure they comply with federal or state laws. For example, if the government or a company wants to build a structure on undeveloped land (e.g., wind farms, solar panels, new buildings) or they want to extract resources (e.g., mining, oil and gas), they will need to have an environmental impact survey conducted. The survey will determine if there are any protected species or the possibility of protected species (i.e., favorable habitat for the species) on the land.

Consultants can also act as a third party or independent source to validate findings done internally by an organization. For instance, if a nonprofit acquires new land they want to survey.

I have never worked with a consulting agency, but from what I've seen from looking at advertised positions, is that there is a lot of emphasis on vegetation surveys. Even if you are a wildlife biologist, you might need to have excellent skills in identifying plants. For instance, you may not just be assessing the presence or absence of species of conservation concern, but if the habitat is also suitable for those species.

Consultants also have to analyze data and write technical reports after they conduct a study. Much of the work that a consultant does has to do with federal and state laws, such as advising clients, obtaining permits, and adhering to environmental compliance rules. Wildlife consultants usually do not publish their research in peer-reviewed journals, but some do.

There are also consulting positions in data analysis and/or with biological samples. For example, when I interviewed for the social scientist position at NCWRC, the previous scientist told me that if he was assigned too many projects and didn't have enough time to analyze the results, he would arrange to have them done by a consulting agency. I've also heard of scientists sending their genetic samples to labs to have them analyzed instead of running the samples themselves or partnering with another professor who does genetics research.

You do not need a higher degree (master's or Ph.D.) to get a permanent position in consulting, and you may be able to do fieldwork throughout your career. Because projects are based on contracts from other organizations, you may have less job security in consulting as companies may downsize when fewer contracted work is secured.

***Examples of Job Types:***
Wildlife biologist, wildlife technician, field biologist, senior biologist, lead biologist, project manager

***Where:***
Private consulting firms

**Need More Help?**

After reading all of these job descriptions and the types of places that you can work, you still might be unsure about what you want to do. That's 100% okay. Although I am emphasizing that you need to know what you ultimately want to do, what's really important is that you have an understanding of the full spectrum of jobs available to you within wildlife biology now. You should be able to cross off a few job categories that you are not interested in. As you progress in your journey, you will likely cross of more. For the remaining categories, you can make sure you get the requirements for those jobs so you will be qualified when you finally decide.

There are other ways to help narrow down your choices later on. In the next two chapters, we'll discuss how you can get experience at any age. These experiences will help you find out what you like and don't like in the field of wildlife biology. Additionally, I am creating a workbook to help you determine the right job for you. Just keep checking fancyscientist.com for updates.

But first, we are going to go over the educational requirements expected of all the different job types. Frequently students are eager to enroll in degree programs, but as you learned from my journey, your degree can overqualify you for jobs.

# 8 DECIDE YOUR EDUCATION

Now that we have gone through all of the major job categories, types of jobs out there, and where you can find work, you should have a better idea of what kind of education you will need. If you used the job tracker, referring to the jobs you've collected will be incredibly helpful. Do you see a pattern with the types of jobs you've listed?

Now let's go over the different degrees you may need to get you those jobs. You can become a wildlife biologist with either a bachelor's, master's, or a Ph.D. degree, but the type of work you will do on a daily basis will vary according to your level of education. You'll want to be prudent here because if you don't carefully research the different career options first, you may end up getting a degree that you don't need. This can eat up your time and therefore money (in many graduate programs you get a stipend), but, more importantly, overqualify you for jobs. We will talk about this specifically later on, towards the end of this chapter. But first, let's start learning about the different degrees and what you can do with them.

**Bachelor's Degrees: Arts or Science?**

You pretty much need a college degree if you want to go into wildlife biology, so this means you will have to decide between a Bachelor of Arts (BA) or Science (BS) degree. This one is kind of a no-brainer; if you want to become a wildlife biologist and therefore scientist, you should get a degree with the word science in it.

But don't fret if you don't. In my experience, this is not that big of a deal. In fact, I have a BA and I made it all the way to a Ph.D. The difference between a BS degree and a BA is that for the BS degree you have to take more science and/or STEM courses.

Why did I get a BA if getting a BS is obviously better for a career in wildlife biology? Remember that when I was in college, I was initially a dual major in theater and biology because I wanted to become a famous actor. Biology was my "safe" choice so that I could become a medical doctor if acting didn't work out.

When I finally decided to become a wildlife biologist, it was after my junior year. By that point, I was close to graduating, and getting the classes necessary to get a BS would have meant that I would need to delay graduation and stay in college longer. To get a BS, I needed more science classes. Since our university did not offer many in the field of wildlife biology or ecology, I decided it was not worth it for me to spend extra time and money in school. Instead, I graduated and concentrated on getting experience through internships. This has never hurt me, and other colleagues of mine also have BAs. I posed this question to Twitter and got similar answers from other wildlife biologists. Getting experience is far more critical.

My opinion is that a bachelor's degree (BA or BS) is not enough to get a permanent job in wildlife biology unless you are willing to move around a lot for the first few years to gain experience. This can even happen to those who graduate with their master's degree too. These jobs may be in remote field locations, usually requiring big moves, and in my experience, it is rare to find temporary positions that line up perfectly in the same city.

While moving around a lot seems fun at first, after several years, it gets old and is expensive. You may also have to have several jobs while working in your full-time temporary position to make enough money to live off of as these positions often don't pay well.

Even after having a lot of experience, it can still be tough to get a well-paying permanent job at the bachelor's level. I do not have personal experience in looking for a permanent job at this level because I only focused on looking for temporary positions (e.g., internships, technician positions) when I graduated from college. As I've scanned the job boards over the years looking for permanent positions now post-Ph.D., it seems like most jobs require at least a master's degree or a bachelor's degree combined with several years of experience. To validate my suspicion, I again turned to Twitter, and also posted to my Instagram stories, and got similar feedback. It's not impossible, but it's hard to do and requires a lot of networking.

The best places to seek employment for jobs at the bachelor's level seem to be the government (state or federal) and in consulting agencies. If you want to work in a field more tangential to wildlife biology, like communication or education, you can absolutely find jobs with a bachelor's degree. However, you will have to research this thoroughly because for these positions, having a degree in fields outside of wildlife biology may be necessary.

## Master's Degree vs Ph.D.

If you've decided to go to graduate school, one of the most important questions you will ask yourself is, "Should I get a master's or a Ph.D.?"

Some of you might be surprised that I am even asking this question. Doesn't everyone get a master's degree first? Then you can go on and decide later if you want to get a Ph.D.? This is the way it has been done historically, but within the past decade or so, it's become increasingly popular for students to just jump right into Ph.D. programs without a master's degree first. As you will remember from Chapter 1, this is what I did.

If you remember my experience, I was already in the thick of my Ph.D., and finishing up for a master's would have taken almost the same amount of time and work as it did for my Ph.D. I was a late bloomer in my research and had not even start collecting data until my third year in graduate school. I also worked on all chapters of my dissertation equally during graduate school, so I couldn't easily drop a chapter without losing the hours of work that I had already put into it. Ultimately, finishing the rest of my Ph.D. would take me about the same amount of time as completing a master's degree, so I went ahead and finished my Ph.D. This is why it is so important for you to know ahead of time the jobs that you want so you can get the proper education to match them.

I think the more significant decision you need to consider is if you should get a Ph.D. at all. In other words, is a Ph.D. necessary for the job you want?

Getting a Ph.D. can overqualify you for many jobs. If you apply for positions that only require a master's degree with a Ph.D., you are likely at a disadvantage. This comes as a surprise to many people (including myself) because you would think that the organization would want to hire the more experienced candidate. However, because the job market is so competitive right now, some people may take any position just to have work and pay the bills. In a 2019 *Nature Careers* article, labor economist Paula Stephan said, "The supply of Ph.D. students was outstripping demand...We are definitely producing many more Ph.Ds. than there is demand for them in research positions."[4]

If employers review an applicant with a Ph.D. for a master's level position, they may think that the person with the Ph.D. is just using the master's level job temporarily until they find a better job elsewhere. I've applied for some jobs that only require a master's degree because I am genuinely interested in these jobs, but I've had to really emphasize in the cover letters why I wanted to work there to convince the employers that this is a job I wanted.

I have generally had less success with these positions and once an interviewer said to me, "You have the C.V. of an assistant professor. Why do

---

[4] Gould, J. *Nature Careers*. https://www.nature.com/articles/d41586-019-03439-x

you want this job?" In some cases, I think I was not invited for an interview because I was overqualified. I know of one case where this was the main reason.

I used to think that a Ph.D. was a "blanket" degree that could be applied to many different jobs within the context of wildlife or conservation. I believe that used to be the case, but no longer anymore given the competitive job market. For example, I met a scientist who worked in wildlife trafficking, which is a more policy-focused and less research-based field. They had a Ph.D. in ecology and the research they did for their dissertation did not directly apply to wildlife trafficking, but of course it did demonstrate that they were competent. However, if they applied to the same position with the same experience today, I don't know if they would get it. I suspect employers would have chosen someone who directly researched wildlife trafficking, and potentially not even a researcher at all. They might have chosen someone with a policy background.

Another thing you have to be careful about is that if you get a Ph.D., it will be harder to find a job where you are outside more frequently. A lot of people contact me saying they want a job like mine because they love being outdoors. Ironically, if you want a career working outside all of the time, I highly recommend that you don't get a Ph.D. In most jobs at the Ph.D. level, you are in an office writing papers and analyzing data. When you get your Ph.D., you now have expertise in designing and guiding research. Your time is better spent administering projects, analyzing data, and writing publications and reports.

Fieldwork is more easily taught. If you are a professor, the fieldwork for your research is mostly done by your graduate students. In nonprofits or the government, fieldwork is frequently done by your staff, technicians, and field assistants. Getting a Ph.D. will most likely overqualify you for a lot of field-based jobs.

Both degrees, a master's and Ph.D. require original research. In a master's program, you will have to complete at least two chapters for your thesis, and for a Ph.D., you will need at least three for a dissertation. Chapters are like subprojects within your larger research question, but they should also stand alone as separate studies to be published in peer-reviewed scientific journals. As an example, my dissertation was on the social structure of African forest elephants. My three chapters focused on: (1) forest elephant home ranges and overlap, (2) forest elephant networks from observational data, and (3) forest elephant networks inferred from spatial genetic sampling. Each chapter resulted in a separate publication.

In a master's program, you are expected to finish faster, around two to three years, because you have fewer chapters compared to a Ph.D. program. Because master's projects have to be done in a shorter time frame, students need to "hit the ground running" when they start. Therefore, their questions are usually developed ahead of time by their advisor.

Ph.D. programs usually take five to six years, but sometimes as short as four or as long as seven years. As a Ph.D. student, you may also come to work on a specific project, but you should formulate at least one chapter/question on your own. Ph.D. students that are not brought in on specific projects (like I was) can usually choose to study what they want within the subject area of their advisor's research and provided appropriate funding for the project.

| Career Type | BA/BS | Master's degree | Ph.D. | Other degrees |
|---|---|---|---|---|
| Researcher | Yes | For many positions, yes | For some positions, yes | |
| Professor | Yes | Yes | Yes, except for community colleges | |
| Educator | Yes | No, but it might help | No | Education, environmental education |
| Communications | Yes | Possibly | No | Communication, business |
| Creative | Yes, unless you are freelance | No | No | Your medium (video, writing, etc.) |
| Law/Policy | Yes | Probably | For some positions, yes | Law, policy |

*The types of degrees you are likely to need for each career type.*

The pros of getting a Ph.D. are that it will open up higher-level positions to you and allow you to choose the direction of the research you conduct. For example, it is impossible to get a job in a university as a professor without a Ph.D. A lot of higher-level research positions at nonprofits also require a Ph.D. Again, you'll just have to do the research on the types of jobs that you want and see what degrees they require.

Those are the major differences between the programs in graduate school, but you should be more concerned about the differences afterward in the workplace. Positions requiring a Ph.D. usually mean that you will be the principal investigator (PI) directing the research and deciding what types of projects to pursue. Within your job, you may have some overarching direction, for instance, a species of conservation concern or a region important to the organization you work for, but there is generally more freedom to choose your research questions and projects. It's commonly expected that you will have to write grants to fund your projects and you will likely be managing other scientists and technicians.

In jobs at the master's level, you will usually be working under a more senior scientist. While you may be involved in grants or research decisions, you will likely not be responsible for them. Jobs at the master's level are more likely to have more time in the field (although this will vary a lot according to the specific position you have), while Ph.D. level jobs are more likely to be in the office managing data, papers, and people. These are generalizations, and there are always exceptions to the rule, which is why it's important for you to search for the jobs you want now to start getting familiar with their requirements.

**Online Degrees**

I get asked about online degrees quite often, mostly if they are worth it and if employers take them seriously. The truth is I don't completely know. If you want a career in research, then you should definitely get a master's in a traditional research program (I don't think there are online Ph.D. programs). For other careers though, I am not sure. Online degrees are becoming increasing common and I suggest you contact employers directly and gather some data as to how online degrees are perceived.

**Other Coursework and Experiences**

While you are in school, pay attention to what other classes you may need that might not be so obvious to your advisor, department, or other students. Use your job tracker to guide you, but here are some broad suggestions (not exhaustive) for some of the job categories and workplaces outlined in Chapter 6:

*Government*:
- Definitely, botany! All of the federal government jobs that I've been interested in require you to take nine credits in botany, a course I never enrolled in. I didn't know this class was important when I was in college and during graduate school, I felt too old to be in an undergraduate class and just felt like I didn't have the time. This was a mistake. Even though I have a Ph.D., I struggled to demonstrate that my research experiences were comparable to a course in botany for the jobs I applied for. I have never been asked to interview for a government position.

- You may also need specific field experiences such as vegetation sampling or specific trapping experiences (e.g., small mammal trapping). Make sure you get those experiences through a field tech position or similar job, or in your undergraduate coursework. You don't need extensive experience, but enough that you understand how to do it and would be able do it yourself with little further training. For example, small mammal trapping is conducted four times a year at the NCMNS and the curator is always looking for volunteers to help check the live traps.
- Many government jobs also desire or require employees to have experience with ArcGIS®. There are even entire jobs dedicated to it if it is something you really love.
- Finally, you may want to take a public speaking course or sign up for free group that you can practice in like Toastmasters. My friend who works in the government told me this is a very valuable skill to have as you are often interacting with the public.
- For research careers, you should join The Wildlife Society's (TWS) local chapter and think about getting certification from their program. Some jobs in the government even require this certification.

## *Science Communication*:
- If you are at all interested in science communication positions, I highly recommend you take courses in communication at your university. Most of the positions that I have seen for science communicators often prioritize communication skills above scientific knowledge. They often advertise for candidates that have degrees in communication, and an interest or secondary expertise in science.
- At the very least take advantage of science communication workshops or seminars offered at your university or scientific conferences. There are many national programs that visit campuses across the country offering these types of trainings.

## *Nonprofits:*
- If you scan job advertisements within specific nonprofit organizations (e.g., WWF's career page), most of the jobs are in marketing, fundraising, development, and policy, and not in research. If conservation is your primary motivation for becoming a wildlife biologist, you may want to prepare yourself for non-research positions such as these by taking courses and/or getting experience in these areas. I, for one, wish I had taken policy and fundraising

- courses and volunteered for a nonprofit during graduate school on a fundraising campaign. My lack of fundraising experience specifically cost me a job as an executive director of a conservation organization that I would have loved to have worked for.
- ArcGIS® experience is often important for nonprofits too.

*Zoos, Aquariums, and Museums:*
- Public speaking skills and other basic science communication skills are important for jobs at these institutions. Even if you apply for a research position, more and more zoos and aquariums are requiring scientists to be accessible to the public through talks and participating in educational events.

*Academia:*
- If you want to go into academia, make sure you get plenty of experience as a TA so you can develop your teaching skills as a professor. This is especially true if you want to become a professor at a R2 university as these universities emphasize teaching. For your job applications, you will have to write teaching philosophy statements and you will want to draw upon real experiences with your students.
- Additionally, check to see if your university offers science education classes, courses on how to teach science. One course at my university was popular, and I know the students enrolled in it found the information valuable and helpful in landing them a faculty job.

## Am I Too Old?

Often people in their 30s, 40s, and even 50s send me messages asking if they are too old to start a career in wildlife biology, especially to go to graduate school. They didn't figure out until a lot later that this is what they want to do and are looking to change their career trajectory. The short answer is no; you are never too old.

At my university in the Ecology, Evolution, and Behavior section within Biological Sciences, there were very few students who went to graduate school immediately after graduating college. Most students were in their late 20s or early 30s, and there were even a few in their 40s. The same was true for the Fisheries & Wildlife department. You can go to graduate school at any time and age you want. Going when you are older and with some experience under your belt can be helpful in providing you direction for your research topic and therefore career path.

That being said, there are financial costs to consider when you decide to attend graduate school later on in your life. You lose the ability to make a decent-paying salary temporarily in graduate school. When you do graduate and get an entry level job with a master's or Ph.D., you won't have as large of a starting salary compared to other careers. Your starting salary sets the precedent for any job after that. You will also have less time to make a higher salary unless you plan on retiring later.

**Financial Considerations**

A realistic consideration to think about before going to graduate school is if you can afford it. Research-based advanced degrees in science are more affordable than other programs because in science, you should get a stipend. That's right; you get paid to go to graduate school! A lot of people don't know that, but as a student, you either are a TA (teaching assistant) or RA (research assistant). You are therefore paid a stipend to teach the laboratory section of a professor's course or conduct research, respectively.

Another perk of graduate school for scientists is that the school usually waives your tuition, so graduate school itself shouldn't cost you anything. At my university, we were also provided with health insurance, however, a few years after I left, the university was no longer doing this.

That being said, the reason why I brought up financial concerns is because you don't get paid a lot during graduate school, and for some programs, they pay you so little that you might have to get a part-time job or take out loans. The Biological Sciences department paid us well, especially if you had the life sciences fellowship, which provided a USD $21,000 a year stipend. This was more than enough to live off of in a small college town.

However, students in the Fisheries and Wildlife department were getting paid far less (I believe around $12-15,000 a year) despite doing similar research. In the biology department, students could easily survive and even save money without having to get extra jobs, while the Fisheries and Wildlife department students could not. Most students skirted the poverty line. On Twitter, one graduate student from another university said she enrolled in the food stamps program so she could afford to eat.

It's also important to think about the loss of time in terms of your overall financial plan. Even though I wasn't going into debt and I was making money during graduate school, the Ph.D. experience cost me almost seven years of my career. Theoretically, I could have been working those seven years and increasing my salary through raises or changing jobs.

No one goes into wildlife biology to make a lot of money and I was always okay with knowing I wouldn't make much. Years later though, I do have to say, it's frustrating how little you are paid given your experience. Since I graduated, things are now more expensive, but salaries have not increased. Unfortunately, now I think one needs to consider finances before going into this career. We'll talk about this more in-depth in Chapter 10.

# 9 GETTING VALUABLE EXPERIENCE AT ANY AGE

There is more to preparing for a career in wildlife biology than your degrees. Internships, technician/field positions, and volunteer work will play a significant role in getting jobs and/or into graduate school. Good work and volunteer experiences in this field will also offer you insight on what type of research you like and want to pursue, as well as what you don't like.

Knowing what you don't like to do is equally valuable information. You may even find out that you don't like research at all. I truly didn't understand research until I went through it myself in graduate school. You may find out that your expectations don't match reality.

Most of your experiences will likely take place during and after college. Still, I wanted to start with pre-college opportunities because I talk to a lot of K-12 students who are already thinking about becoming wildlife biologists. Many of my colleagues said they knew they wanted to become a scientist from a very young age. Additionally, many parents contact me for advice because their child wants to become a wildlife biologist, even from ages as young as four, and ask me how they can help prepare them. Therefore, this section is also written for parents who have small children too young to read this book yet, but they still want to foster their interest in wildlife biology.

If you are older, I still suggest that you read the before and during college sections because there are some opportunities that you can do at all levels (e.g., citizen science) that will not be repeated in later sections.

**Scientific Mindset:**

It is the job of a scientist to ask and answer questions about the natural world. Fostering a sense of curiosity and skepticism in your child will serve them well if they decide to become a scientist. If you are an adult or teenager wishing to become a scientist, tap into your inner younger child and try to look at the world with fresh eyes, questioning everything.

When kids are young, it's essential to foster and maintain their innate sense of curiosity because over time, as a society, we often teach them not to be. Kids who ask a lot of questions often get in trouble. When I was in school, we were told there was one right answer. This taught me to stop thinking once I knew what the answer was and memorize that correct answer for the test.

One of the most important things you can do for your child is to keep them curious. Reward them for asking good questions through positive feedback and encourage them to ask more questions about the natural world. Investigate, research, and explore the answers together. Even if you know the answer, allow them to explore the wildest possibilities before finding out or telling them the correct one.

During graduate school, I was shocked to learn that there are still many things about the natural world to be discovered, things that we don't understand, even in our own communities. I always imagined such discoveries were far away in remote, inaccessible rainforests but this is not true. It is the job of a scientist to ask and answer questions about the natural world. Fostering a sense of curiosity and skepticism in your child will serve them well if they decide to become a scientist.

In addition to developing a scientist's mindset, here are some other, more concrete ways you can involve kids in wildlife science and help prepare them for such a career.

## K-8th Grade (Elementary and Middle School)

### *Nature and STEM Afterschool Clubs/Camps*:

Girl scouts, boy scouts, 4H, or any other club that gets kids outside will help foster a child's interest in wildlife biology. But you don't have to just limit their experiences to clubs based outdoors; getting your child involved in science clubs overall can make a difference. Remember, wildlife biology is really all about science.

People have this perception that wildlife biologists go for walks in the woods and take an inventory of all of the plants and animals that they find, taking pictures and making drawings of all of the different species. This is more the work of a naturalist, which quite honestly, doesn't really exist anymore as a career. While wildlife biologists may conduct a survey in the woods and have to identify different species as part of the methods, this process is done within the context of a robust study design. To be a wildlife biologist, you have to have a strong understanding of science.

There is a big push to get kids interested in STEM fields because over the next few decades, more and more jobs will be created in many sectors within STEM careers. However, there are not enough people projected to have the skills necessary to fulfill those jobs. The National Science Foundation allocates millions of dollars each year to researchers and organizations across the United States to develop programs that encourages and increases children's access to STEM fields. A lot of emphasis is placed on traditionally underrepresented groups such as minorities and females.

*Citizen/Community Science*:
A great approach to teaching kids about science is by allowing them to do real science. This is now possible through citizen science (also called community science). Citizen science programs consist of real scientific research that members of the public can participate in.

I can tell you from experience that in many of these citizen science programs, you are doing the same work the scientists would be doing. For example, in eMammal, I have to expert review hundreds to thousands of camera trap photos nearly every month to ensure that the species identifications by volunteers are correct. The volunteers set up the cameras, upload the photos, and identify the species, just like we would.

When a camera trap project accumulates many photos, it becomes too much work for the scientists to do alone. For example, in Snapshot Serengeti, a camera trap project focused originally on understanding lions and their prey in the Serengeti ecosystem, cameras collected millions of photos over the course of the project. Although the focus was on lions, there are many animals that live in the Serengeti that triggered the camera traps. These other "by-catch" animal photos are still valuable for conservation and science. However, there were just too many for scientists to do alone to finish the project on a timely basis. Therefore, the scientists recruited thousands of volunteers to help them identify the species in the photos just as they would.

If mammals are not your thing, you can find citizen science projects of any kind through SciStarter.org. Scistarter is a gigantic database containing citizen science projects all over the world. On the website, you can search for projects that align with your area of interest (e.g., birds, astronomy, genetics, etc.) and/or your lifestyle (e.g., remote, virtual projects or outdoor hands-on ones).

Citizen science programs vary greatly. Some are local, and you may have the opportunity to meet the scientists on the project and even work alongside them. For example, I read about one project where participants tracked urban hedgehogs alongside scientists. Others are international and much more independent. For example, iNaturalist can be done anywhere around the world with just a smart phone to take photos and upload data.

## Go Outside and Explore:

This is a big one, but also super easy to do. Ask any wildlife biologist why they became a one, and they will say it's because of their experiences being outside as a child. I've only met one person thus far who has not answered in this way, but their response was close; they said watching wildlife on television when they were young was what inspired them to become a wildlife biologist.

The outdoor area that you explore does not have to be pristine or far away. You don't have to go to Yellowstone National Park or drive an hour to a wildlife reserve. It can be as simple as going in your front or backyard or a local park. When I was a child, one of my favorite things to do with my family was turn over rocks in our suburban front yard to look at all of the insect life just below our feet that we couldn't see. I felt like it opened up to a whole other world and under every rock was a different surprise.

There is still plenty left to discover in this world and there are living organisms everywhere, even in the most urban cities. You just need to pay closer attention and assume that not everything has already been figured out. One professor I worked with had a study on ant diversity in road medians in New York City. There really is nature everywhere.

If you struggle with figuring out what to do outside, you can again turn to citizen science. Find a program that takes place outdoors, so you have directed and purposeful activities that also contribute to scientific research. Citizen science programs teach you to closely observe the natural world and may even give you some really beneficial skills for a career in wildlife biology.

You can also work on your identification skills for local species or particular taxonomic groups. Someone who can readily identify vegetation, birds, and other species will likely be more competitive for tech positions and even some jobs post graduate school.

## Consume Wildlife Media:

This may sound like a lazy way to learn about wildlife, but there truly is excellent programming out there that teaches you about science and nature, and in a captivating way. Honestly, I even learn a lot when I watch some of these shows. There are so many crazy, beautiful, awe-inspiring species on this planet and it's impossible for one person to know about all of them. The BBC's series *Planet Earth* and *Blue Planet* are especially noteworthy docuseries that foster curiosity and teach essential scientific principles like evolution. I grew up reading National Geographic for Kids books and looked forward to my monthly *Ranger Rick* magazine. To this day, I can still remember spending hours with my dad reading these books over and over again.

You will have to be careful about what programming to watch though, especially in this Internet and social media era. Just because a network or a YouTube channel seems educational does not mean it is science-based.

Scientists have complained about national networks featuring shows with obvious misinformation (e.g., mermaids), but making them seem scientific misleading the public.

You especially have to be cautious on YouTube as this is completely unregulated. People can say whatever they want, and claim they are experts when they are not. In general, it's good to look for channels where the hosts have higher degrees (master's or Ph.D.) as they will have a better understanding of the scientific process and the subject matter more in-depth.

## *Coding Classes/Clubs:*

Coding deserves a separate nod from STEM clubs because most people would not associate coding with wildlife biology, yet it is extremely important in research. When people think of wildlife biology, they automatically think of outdoor activities like birdwatching or tracking animals. However, coding is an essential skill.

With the rise of big data because of technology and increased computing power, analyses for research are becoming more statistically complex. Wildlife biologists write their own, or at the very least, tweak existing code for the statistical analyses and models that they run. Knowing how to code at a young age would offer a considerable advantage to anyone who wants to go into this field.

Wildlife biologists mostly use the statistical program R® but learning how to code in general will get you into the right mindset for coding in other computer programs. It's kind of like learning a new language; when you learn one young, especially at a young age, learning another one later on in life becomes easier.

## 9th – 12th Grade (High School)

In addition to the suggestions above (many still apply to high school kids), you can also add in these activities:

### *Science Elective Courses*:

Take any science elective that you can in high school. It doesn't matter if it is related specifically to wildlife biology. The basic sciences (e.g., chemistry, physics, etc.) are all required for wildlife biology degrees at the university level, and there can be intersectionality between wildlife biology and other sciences.

### *Advanced Math Courses*:

I took calculus in high school, and I specifically remember thinking, "When am I ever going to use this stuff?" Well, it turns out, now. In lab meetings, we often read papers that involve calculus.

Again, because wildlife biology involves data, and therefore data analysis, knowing advanced math and statistics will serve you extremely well in this career. Take as many advanced math courses as you can, especially in statistics. In the online course, I've also included books you can buy to learn statistics and R® on your own.

***Volunteer in a Structured Program:***
When you become older, there may be opportunities for you to volunteer. Volunteering is a great way to obtain experience if you can afford to donate your time (more on this later). The opportunities will vary by institution, so you will have to do research and maybe get a little creative.

Look at institutions like zoos, aquariums, and museums to see if they allow volunteers younger than 18. For example, NCMNS had a junior curator program. I read recently in one Facebook group that I am a part of, that someone's intern had been volunteering since they were 13 years old!

At this stage and because your opportunities will likely be limited, do not be choosy. Any volunteer program that involves wildlife will be helpful, even those that aren't technically related to careers in wildlife biology (e.g., volunteering at a wildlife sanctuary, zookeeper work).

***Volunteer at a Lab:***
This suggestion is for the more ambitious student, and it may be hard to acquire even if you really want the experience and are an excellent candidate for it. Professors or their graduate students may be unwilling to supervise a high school student, and some institutions may even restrict volunteers to be 18 years and older.

That being said, it doesn't hurt try! One of my recent undergraduate researchers volunteered in a lab when she was in high school and I mentored a high school student as a graduate student, so this is something that you can make happen. In fact, at eMammal, I even mentored some middle school students that wanted to help out with uploading photos.

What I mean by volunteering at a lab is to find a local research lab or university where you can help the professor and/or the graduate students with various tasks. To do this, identify universities located near you, and for each university, search the appropriate departments on the university's website to find individual professors' research pages. The professors' websites will provide you with a summary of the type of research that they do and the titles of their peer-reviewed publications. Do not limit your search to wildlife biology departments; many departments include professors that conduct wildlife research, but they are listed under different names such as biology or ecology. Remember that I volunteered in a flour beetle lab studying sexual selection within the Biological Sciences department at my university. Any research experience in science no matter how unrelated to wildlife biology, is more valuable than none at all.

After you find a professor whose research interests you, contact the professor through a professional email asking if you can volunteer in their lab. Do not have your parents write an email for you. Doing it yourself demonstrates that you will be mature and responsible enough to handle a volunteer position. See the online course for an email template.

Because you are in high school, you will probably only be able to volunteer during summer months when you don't have school and can devote spending more time in the lab. Some researchers do not come into the lab on the weekends, and an hour or so after school is likely not enough time to volunteer, especially if you are doing lab or fieldwork.

Be willing to do tedious chores like cleaning glassware, sorting field equipment, and data entry. You have to start somewhere and all of us do these activities no matter how educated you are. Volunteering experiences are also not necessarily about what you do, but more about the knowledge you will gain from interacting with the professionals you volunteer with (professors, graduate students, and postdocs in the lab) and the network you will build. Getting your first reference and your first experience in science to put on your resume is key for future jobs. The earlier you can do this, the easier it will be for you to get jobs later on.

**During and After College**

This list is similar to the types of experiences you can have in high school, but the experiences will be at a more advanced level. Even if you are in college, remember that you can still do activities from the K-12 list. You can learn a lot from citizen science throughout your entire career.

Before we begin, although these sections may have different titles, essentially, they can be interchangeable. In my view, volunteer, intern, field assistant, and technician are mostly differentiated by how much pay you get (with volunteer positions being unpaid) and whether or not you are in a structured program (internships).

There are some nuances, which I will discuss, but these nuances do not reflect whether one experience is better than another. An internship is not necessarily better than a volunteer or field assistant position just because it is a more structured program. The worth is determined by what experiences you have in the position, how much you learn, and the connections you make for networking.

*A Note on Economic Disparity and Financial Security with Temporary Positions:*

Many nonprofits, professors, graduate students, and other institutions accept volunteer positions which creates a bias of participation in science towards those who can afford it. If you need to work a job in college to pay your bills, you probably do not have extra time to volunteer. Therefore, you wouldn't be able to get the critical experience you need to get future positions or into graduate school.

Sometimes students not only have to do the work for free but have to pay for their experiences. Crazy, right? I kind of did this for the SFS internship in Kenya. I had to pay for half of my airfare to Kenya, and I was paid in a Kenyan salary. SFS did provide food and housing, but the cost of the airfare and my lower wage in Kenya cancelled out and I made no money that year.

Positions abroad and/or those that include work with charismatic species are so coveted and competitive, it is not uncommon for you to have to pay for your plane ticket to get to the field site and room and board while you are there. These experiences can cost you thousands of dollars for just a few months. This is of course, unfair because there are many students who are equally or even more qualified to participate and deserving of the experience but can't because they cannot afford to pay for them.

On the other hand, it is very challenging for scientists to get funding to pay salaries and scientists could be depriving them of an important research opportunity. I receive emails all the time from students inquiring about volunteer research opportunities in our lab. Volunteer work obviously helps them with their career, and it helps scientists advance their research. With fewer paid positions, fewer people overall will get these experiences, which makes it fairer to those who can't afford to volunteer. In the past six years, I've only been able to create one paid temporary position. The majority of the students that worked for me were volunteers.

Unfortunately, this is currently how science works and although professionals are discussing how to fix it, I don't see the system dramatically changing in the near future. Some scientists refuse to hold unpaid positions in their labs to create equal opportunities, but I'm afraid those professors are few and far between.

There are some things you can look into if you cannot afford to volunteer, but still gain research experience. Universities usually offer course credits for working in a lab. Although you won't be earning money, it will count towards your credits for graduation. Some universities may offer a work-study program where you can work in a lab and are paid by the university. This was the case for NC State University students who worked in our lab. You can also see if there are any scholarship or small grant opportunities at your school or even elsewhere by searching online. If you are a member of an underrepresented group in science (a person of color or a woman), you can look to opportunities outside of science that specifically fund these types of candidates.

Unfortunately, you do have to consider finances before entering this career. To be completely honest, had it not been for my family, I don't think

I would be a wildlife biologist because of the potential risk. I took three internships after college, and while I did get paid for them, my jobs did not line up. I moved back home with my family in between periods of employment. Those who do not have that option will have to immediately get jobs (in or out of this field) to support themselves.

Some positions pay stipends instead of a minimum wage, meaning that you may be working a full-time job, even over 40 hours a week, but are still not making enough to pay your bills. I know many people who work two to three jobs to support their tech jobs in wildlife biology.

Even now, I rely on my husband for financial stability. I've been in a postdoc position, which is temporary and was unable to find work in my field in Raleigh. Since my postdoc ended in February 2020, I transitioned my career to science communication and started my own business (the Fancy Scientist, LLC), but have yet to make enough of an income to support myself.

Some of my colleagues took several postdocs before they secured a permanent position, requiring them to move multiple times. This becomes expensive. Having another position lined up is not guaranteed, and one of my colleagues was unemployed for several months. I know others who had to leave this field completely because they couldn't find work that would pay enough.

### *Volunteer in a Lab:*

In my opinion, volunteering or getting paid to work in a lab is essential experience for an aspiring wildlife biologist. I discussed this previously in the high school section, but I am writing about it again here because it is so important, and I will give you more details as to what it is like at the university level. This experience gives you first-hand knowledge of what it's like to work in research, it starts the growth of your scientific network (who will help you get jobs later on), and it will help you determine what kind of research you like and don't like.

Before we get into how to gain experience, let's discuss what I mean more by volunteering in a lab. I don't mean that you have to physically work in a laboratory with beakers and test tubes (although this is a possibility). By lab, I mean volunteering under a scientist who conducts original research. Universities are the most well-suited for such experiences, although other organizations like zoos and museums will take volunteers.

When you volunteer in a lab, you are volunteering to help the principal investigator (PI; at universities, this will be the professor) conduct their research. Often you will not be working with the PI directly, but a postdoc or a graduate student instead. Graduate students conduct their own research and frequently need volunteers and/or field assistants to help them.

In wildlife biology, the lab space is usually where students have their desks and offices. The real lab space where you collect data is the outdoors unless the data have already been collected. If the lab you work in does have samples that need some sort of processing, though, you may also physically work in a lab space. For instance, in graduate school, I had an office for reading the literature, writing papers, and data analysis, and a lab space to process my genetic samples.

Now that we know what a lab is, how do you go about getting a position in a lab? There are two main ways:

The first is to look for advertised positions and apply for them. You can look on job boards and your university's job board. You will likely have better chances at finding opportunities at your university though by just exploring the halls or asking around. Labs will frequently advertise that they are looking for volunteers or technicians through posted advertisements.

In my opinion, the best way to get volunteer experience is to read about the professor's research you are most interested in and approach them directly by asking if you can volunteer in their lab. Do this either by writing a professional email, or if it is a professor that you know, for example, you are in their class, you can talk to them in person during their office hours.

It is more impressive for you to ask them in person; however, some people are shy, intimidated, or hate to be put on the spot. Email also gives you the advantage of attaching a resume. If yours is impressive, this may sway them. You can always send the professor your resume as a follow-up to your in-person conversation too unless they give you a hard no.

Some labs are very competitive and hard to get into even for volunteering. It is not uncommon for a professor to completely ignore your email, never give you a response (another advantage of asking in person; they can't ignore you), or say no. If you get no response, I would send a follow up email in a week or so. Many professors are overwhelmed with the number of emails that they get; it is easy for yours to get lost in their inbox.

If you still don't get a response from the professor or even a no, you can also try emailing the graduate students that they are advising in their lab. Working under a graduate student does not mean your experience will be any less valuable, and even if you do work with a professor, frequently you will end up working under one of their graduate students. Working closely with a graduate student gives you the added benefit of being able to ask them a lot of questions about graduate school and they may be more approachable.

If you do work with a graduate student, make sure you still establish a relationship with the professor advising them. When you apply for graduate school, and if you apply for the National Science Foundation's Graduate Research Fellowship Program, you will need letters of recommendation. A letter from a graduate student is not as compelling.

You can create stronger relationship with the PI by asking to sit in on lab meetings where students give updates and/or read scientific papers as a group. If you are invited to a social function such as dinner with the lab or an event at the advisor's house, always go unless you have a conflict you cannot reschedule. These are great opportunities for the professor to get to know you.

If you get a hard "no" from professors or students, don't panic or be devastated. It's more important that you start somewhere in any lab and get experience than no experience at all. You just want to get your foot in the door. Try for labs that you are not as interested in to get some basic experience and remember that you can always ask someone who rejected you previously again the following year. Asking multiple times shows that you are highly interested in their lab and have follow through.

### *Internships*:

Internships are a little bit different from volunteering in that they are supposed to be more structured. Internships are designed to not only provide you with an opportunity in a research setting (or another setting such as education or outreach) but also provide you with professional development. For example, during my Disney internship, we took short courses and I shadowed other scientists to get more experience, even though their research was not directly related to the projects that I was working on.

My internship with the Bureau of Land management was part of a more extensive program with the Chicago Botanic Garden. Before the interns were deployed to our government posts around the country, we met in Chicago where we had several days of training in using GPS and making maps in ArcGIS®. Although internships are supposed to provide more structure, in reality, you may get the same types of experiences that you would get from volunteer or tech positions. It just depends on your mentor and the organization you are working for.

### *Technician (Tech)/Field Assistant/Lab Assistant:*

Technician positions are similar to internships and volunteering, except that these positions are paid (although internships can be paid too). Technician positions are common both in university research labs and in the government. Less frequently, I see technician positions advertised for nonprofits.

If the nonprofit is international, these positions are usually designated as field assistants, and the nonprofit will likely hire locals. Having lived in country for their entire lives, locals are more knowledgeable about the people and wildlife there. Employment of the local community also empowers them and can even mitigate conservation problems. For instance, nonprofits will sometimes hire former poachers who initially turned to poaching because they needed money but now can use their same skill set to protect vulnerable species while having a stable salary.

In the US, tech positions are almost always temporary. Because you are being hired for specific work, less emphasis will probably be placed on your professional development. However, this entirely depends on your employer and how much they care about mentoring you. If you show initiative, drive, and dedication by asking great questions (at the right time, not when they are busy), they will be more likely to mentor you even if that is not their natural inclination.

A lot of technician positions in wildlife biology are in fieldwork. There are some amazing opportunities that allow you to travel to remote areas of the world. Fieldwork is often people's favorite part of wildlife biology, but reality does not always match expectations. You may face adverse field conditions like swarms of mosquitos, extreme heat, and monkeys pooping on your head (I've seen this listed on postings before). Unfortunately, there are also real threats in participating in fieldwork, especially for women.

Within the past few years, studies have shown that sexual harassment is not uncommon in the sciences. Many women have started to speak about their own experiences on social media. When you are in the field, you are frequently with small groups of people and sometimes only one other person for extended periods of time. In my Ph.D. work in Gabon, I would spend up to eight hours a day with a male field assistant alone. My field assistants were fantastic, extremely professional, and they quite literally saved my life. One prevented me from getting trampled by an elephant that had snuck up on us while we were collecting a dung sample and charged us severely. While I have never had any experience even remotely close to being sexually harassed or raped in the field, it unfortunately does happen.

## Ways to Level Up Your Experiences

### *The Importance of Publications*

If your ultimate career goal is to secure a research-based position, during your volunteer/internship/technician experience, express your interest in contributing to a manuscript if it is possible and appropriate. I was really lucky that the wildlife management professor at SFS approached me about helping him with several papers when I was an intern. I ended up getting two publications from this internship, and one of them is still my most cited paper to date.

When you approach this subject with the person mentoring you, make sure you do not go about it as if you are entitled to a publication. To be a coauthor on a paper, you have to contribute significantly to the original research you are working on in one or more forms: in the concept/design, data collection and analysis, and/or writing. Collecting data alone is not enough for a person to be added as a coauthor to a paper, especially if you are being paid to do so. You need to contribute intellectually. For the papers that I was a coauthor on in Kenya, I helped collect data, manage the students on the research team, analyzed the data, and wrote early drafts of the manuscript.

### *Maximize Long Car Rides*

Long car rides may sound like a bizarre topic to bring up in a section on professional development. What do car rides have to do with wildlife biology? Because wildlife biology at the volunteering/tech/internship level almost always involves fieldwork, you will likely have the opportunity to travel with people more experienced than you to field sites. This could be the PI or the graduate student you are working under, but it could also be other people that have been working in the field for a long time.

These extended car rides are an excellent opportunity for you. In Utah, I rode with my boss in the car for hours to our field site near the Grand Canyon, and in Kenya, almost every place we visited was at least an hour away. Even in graduate school, I was a tech on an amphibian project, and I asked the more senior graduate student for her advice and perspectives for the hour-long car ride we had each way.

Take advantage of these opportunities. You are getting uninterrupted time with someone who is way more knowledgeable than you about this field. If you were to contact a professor and ask them if you could interview them for even thirty minutes, most would likely decline, unless you were in their class and/or in office hours. But until driverless cars come along, they will likely be the ones driving and therefore available for useful discussions. Ask them questions about their career path, their research, their students' projects, and what it's like to work where they do. Asking genuine questions will help you learn a lot about this field and people love to talk about themselves. Many of my career decisions were a result of discussions from long car rides.

### *Maintain a Good Attitude*

To many people's surprise, introductory work in this field is often not much fun. It's likely to be tedious and repetitive even if you are working in a cool place or on a flashy project. This, unfortunately, is part of science. In order to answer scientific questions, we need to collect lots and lots of data. This often means doing the same thing over and over again. Even with a Ph.D., I still have to do repetitive tasks like reviewing camera trap photos or tedious ones like data clean-up. While reviewing photos can be really fun and is one of my favorite parts of my research, there are deployments that have thousands of photos of humans, cows, or sheep. Those deployments are not fun to review.

As a volunteer or tech in a lab, you may have to enter data, clean up data, watch videos of data, or sort through photos. If you are in a traditional lab with equipment, you may have to do things like wash dishes or run PCRs and gels over and over again. I ran hundreds of PCRs myself in graduate school even with the help of undergraduate volunteers.

If you are in the field, you are likely going to have to deal with unexpected challenges and discomfort: getting rained on, long days, flat tires, extreme heat or cold, and the absence of fresh food, to name a few. No matter how miserable you are, keep it to yourself and maintain a positive attitude. While it may be super annoying to get a flat tire, it's even more annoying when the people around you are complaining about it.

When employers call your references, they almost always ask about your attitude. Many of my colleagues and I would prefer to hire a mediocre student with a great attitude, compared to a stellar student with a poor attitude. Don't make the mistake of complaining, starting drama, or being difficult. Otherwise, it will be hard for you to get a job in the future. Try not to make demands unless they are absolutely necessary, such as for a health reason.

### *Statistics Courses:*

Once again, a big part of wildlife biology is data analysis. You cannot get enough statistics classes even if they don't seem that relevant to wildlife biology. My universities did not offer good statistics courses. To this day, I struggle more in data analysis and have to work harder than others. Take as many statistics and modeling courses as you can.

### *Writing Courses:*

People don't realize this but reading and writing is a huge part of science and our work as scientists. No matter what your job is, you will likely have to summarize your results in a written report. Learning how to write, especially succinctly and with clarity, will serve you well. You might want to consider taking writing classes during your undergraduate studies.

I recommend journalism courses over English classes. Journalists write nonfiction articles in a narrative form and are often limited by a word count similar to how scientists are limited for manuscripts. Journalists write with the intent to get straight to the story and are not supposed to write with flowery language. Science writing is similar in that regard. Also, if you are interested in science writing as a career, you will already have some credentials.

### *Science Elective Courses (Especially Botany):*

Try to take as many science classes as you can. I personally think it's important to get taxonomy classes, especially botany. Maybe this was because my university didn't offer these types of classes and therefore, I feel deficient in this area. However, I emphasize botany again because through my years of scanning the job boards, it seems like a lot of wildlife jobs desire and even require plant expertise even outside of the requirement in many government jobs. This makes sense because an animal's habitat is critical to their survival and therefore a big part of your job may be evaluating habitat suitability.

## *Conferences:*

Attending regional, national, or international scientific conferences are an experience you will want to have during graduate school, but you can also go to conferences during your undergraduate degree. Professional conferences are where scientists and practitioners go annually (some are done every other year) to discuss the latest research, advancements, and innovations made in a specific area of science. Conferences are also a fantastic place for networking, and the more established you are in your career, the more you realize that is their primary benefit. They are also great opportunities for professional development.

As a student, you will want to think about conferences from two different perspectives. Conferences are events to present your research to professionals in your field and where you can meet people you may want to work with or organizations you may want to work for later on in your career. For instance, you may want to meet with a specific professor as a potential advisor or postdoc advisor. As an undergraduate student, conferences give you the opportunity to meet with potential advisors even before you apply to graduate school.

For graduate students, postdocs are becoming more and more common nowadays even for people who do not want to go into academia. You can attend talks given by those professors and/or their lab members so you can get to know their research and start networking with them.

You can also research different workplaces. Conferences always have vendors set up at tables usually where the poster sessions are. Frequently, nonprofit and government organizations send representatives to talk to people about their institution. Attending conferences will allow you to speak directly to employees of those organizations. You will be able to gather important information on what can make you competitive for a job at such a company. This will help you make sure you get the skills you need while you are still in undergraduate or graduate school. You can also develop professional connections directly with them early on.

My recommendation is to attend as many conferences as you can as long as they align with your career objectives, and you can afford them. If your lab is offering to fund you, then I would take advantage of this opportunity. Sometimes you can attend too many conferences, and this will not only drain you (and your resources) but can slow down your progress towards your degree. Make sure they are appropriate for your research and networking strategy.

In my opinion, the best conferences for wildlife biologists are the Wildlife Society's annual conference and the conferences put on by the Society for

Conservation Biology. They have a North American meeting every other year, the North American Congress of Conservation Biology, alternated with an international one, the International Congress of Conservation Biology. I've found other broad topic conferences (e.g., Ecology, Behavior, Evolution) to be not as focused on wildlife.

I also recommend going to conferences specifically for your taxon. For example, for me, it would be the American Society of Mammalogists. These conferences will often be the best because they are in your niche. I must confess though that I have never been to a Mammal meeting; the timing just never worked out for me.

# 10 GOING AT IT "FROM THE SIDE"

One night after a professional meeting, I went out to dinner with colleagues. Somehow, the conversation shifted to one prominent wildlife biologist telling the rest of us how he was envious of his non-biologist friends traveling the world and having these incredible experiences. This caught me by surprise.

He explained their well-paying jobs gave them the time off and provided enough income to take extravagant vacations regularly to different biodiversity hotspots around the world. As the biologist was naming countries that his friends traveled to, again, to my surprise, he included some countries where he actually does long-term research and has visited regularly.

I interrupted him and pointed out the fact that he has been to and has research projects in those countries. He went on to say jokingly that he never really went to those countries because he was always in meetings in conference rooms. He said his friends showed him photos from all of the gorgeous places they visited within those countries, places he had never been to. Even as a prominent wildlife biologist, he didn't have the time off or excessive income to go hike mountains, see waterfalls, or take safaris for fun.

A lot of people want to become a wildlife biologist for some or all of the following reasons: they want to have a job where they can be outside all of the time, see/be close to animals, conserve wildlife and their habitats, and/or travel to exotic places. However, in reality, wildlife biologists do not necessarily do these things. Or at least to the extent you may think they do. The prominent wildlife biologist I was speaking to spends most of his time in meetings and writing papers inside even if he does have research projects in exotic countries on some of the most charismatic species on the planet.

This wildlife biologist and many others do work on the direct conservation of wildlife and their habitats. However, you would be amazed by how much conservation work does not directly involve wildlife science. Organizations like WWF, WCS, and CI to name a few are nonprofits that of course hire Ph.D. level researchers and use science-based evidence to inform conservation decisions.

That being said, implementing those decisions often doesn't always involve science, or if it does, it involves the social science of human behavior. A lot of time and effort is spent by these nonprofits getting community buy in, in meetings with government officials or other stakeholders, and in educational programs for the community.

Ultimately, most conservation solutions point to us as human beings. We are taking habitat away from species, poaching or overhunting species, changing the climate at a rate that is too fast for species to adapt, and dumping pollutants into their environment that makes them sick. These are the main reasons why much of the world's biodiversity is threatened. The solutions to these problems frequently lie in political and/or economic means and changing human attitudes and behaviors.

In fact, the perpetuating theme throughout our curriculum when I studied abroad in Kenya was the importance of the local community and that conservation would not and could not be successful without their support. We heard stories about how as a strategy to reduce poaching, the Kenyan Wildlife Service protected endangered animals through protected areas, limiting access only to paying tourists. This was a success, however, when animals left the park, they were speared in retaliation by the local people. Most protected areas cannot support wildlife populations alone; they need "spillover" areas outside of the park boundaries for animals to persist. In fact, the majority of wildlife in Kenya is found outside of protected areas. Once high numbers of endangered species like rhinos were being killed, the Kenyan Wildlife Service was forced to rethink their strategy.

Even though these lessons were repeated to me over and over again in my study abroad experience, I didn't really take them to heart until I started studying forest elephants for myself. I wanted to help rescue this species from endangerment by learning about the elephants' intricate social structure, increasing worldwide attention to this species through my research and scientific advancements in knowledge.

However, when the scientific papers that I had spent years writing finally came out, there was little reaction. I felt proud of my scientific accomplishments but was sad that I wasn't doing more for the species that I cared about so much. The following year after I graduated, a new paper by one of my colleagues in Gabon found that between 2002-2011, the duration of my Ph.D. plus a few years, over 60% of the entire forest elephant population declined due to poaching[5]. The poaching was almost exclusively driven by the consumption of their tusks as sources for carving statues,

---

[5] Maisels *et al.* 2013. Devastating Decline of Forest Elephants in Central Africa. *PLoS ONE* 8(3): e59469

jewelry, and other decorative objects.

The true conservation issue had nothing to do with studying the elephants themselves. What was the point of studying a species if it might not exist in a few decades? If I really wanted to help forest elephants, I should have been studying the people, the consumers who were purchasing ivory to determine if there were ways to change attitudes towards ivory and purchasing behavior. Yes, having rangers on the ground to protect parks and elephants is important, but if there is no decrease in demand, it will constantly be an uphill battle. All of the solutions to the conservation problems of forest elephants are social, political, and economic first.

If you are interested in pursuing wildlife biology as a career for conservation purposes (like I was) or because you love animals (also me), you might be better suited in another career if research is not your thing but can still work for a conservation organization. Nonprofits need lawyers, financial planners, fundraising experts, and marketing executives to name a few. When I perused the job boards of nonprofit organizations, I was surprised by how few research positions there were. There were far more in fundraising, marketing, and development.

Even if you don't work directly for conservation, honestly, you can still make a difference and help conservation efforts in other ways outside of your career. A lot of conservation is really about investing in programs and habitat, so species stay protected. For example, if you can purchase and/or donate money to organizations that buy large areas of land, this land can be set aside for wildlife conservation. The biggest threat to wildlife is habitat loss and simply buying more land, keeping it undeveloped, and/or restoring it for species to live on, is one of the major means to solve the biodiversity crisis.

Conservation organizations also have to spend funds to ensure the security of protected areas, even if they are officially designated as national parks by the government. Many parks are called "paper" parks because although park boundaries are drawn in on a map, there is little to no law enforcement to stop poaching or other illegal activities like mining and logging. Donating to nonprofits that invest in rangers to patrol these important protected areas is, therefore, a very effective conservation solution. Quite honestly, this is the best way to protect forest elephants right now. If you earn a larger salary in a more lucrative career and donate a portion of your income to conservation organizations so they can protect wildlife and their habitats, you are still making a big impact.

I also think it's important that we have people who care about conservation and wildlife across all different types of sectors and disciplines of the workforce. If all of the conservation-minded people strive for careers in wildlife and conservation organizations, there will be no people to drive conservation and sustainability efforts across other places of work.

Corporations have a massive impact on wildlife. For instance, how corporations develop land (mining, agriculture, forestry), their carbon footprint, and how they deal with waste and pollution all affect wildlife. People inside these industries who care about wildlife can advocate for more sustainable practices that have conservation in mind. In Gabon, there is an oil company that heavily controls access in and out of their reserve, which inadvertently prevents poachers from entering. This reserve has become a safe place for local elephants. Climate change affects every single animal on this planet, and even if you don't work in conservation, you can impact wildlife by advocating for more sustainable, low carbon practices to be adopted in your corporation.

If you want to go into wildlife biology because you love being outdoors, there are some other considerations you may want to take into account before making it your full-time career. Sometimes when you convert your hobbies into a job, it can take all of the fun out of them. Many people use their hobbies as an escape from work. If what you love to do becomes work, the activity may now be negatively clouded by responsibilities, bureaucracy, a demanding boss, and/or toxic workplace conditions.

Additionally, your job may not offer you the same type of satisfaction as your hobbies do. One person contacted me saying they wanted to quit their job at a software company to turn their love of scuba diving into a career in marine biology. Doing scuba diving as a marine biologist could backfire because the types of experiences that you engage in with scuba diving as a tourist are likely going to be very different from those that you experience as a marine biologist. It's unlikely that you are going to research the beautiful coral reefs that divers are attracted to.

Scuba diving may even lose part of its fun when you have to collect data instead of just exploring. You may also find it challenging to secure the funding necessary to complete your scuba fieldwork as grants are often competitive, and scuba diving is expensive. Rather, if you did scuba as a hobby while working 9-5 at a software company, you might be able to do it more frequently and stress free. I am not saying you should not pursue marine biology as a career, but I do think you need to think through all of these different factors.

Where you travel to for your field research will all depend on your study questions. You may envision wildlife biologists collecting data in a beautiful rainforest or on safari in an African savanna, but a lot of wildlife biologists study species outside of protected areas. Studying species in human-developed landscapes is especially true in conservation biology, as scientists are often trying to understand how animals respond to real-world conditions. For instance, you might travel to Indonesia to study wildlife, but if you are studying the impacts of palm oil on mammals, you may be spending your time in palm oil plantations rather than the gorgeous primary rainforests that tourists visit.

Through citizen science, there are a lot of opportunities that are available for you to participate in as a civilian. If you remember from earlier, citizen science is research that includes non-professionals as part of the research process. Usually, volunteers complete a tutorial or training program and then help scientists collect or analyze data.

In our eMammal program, volunteers set up the camera traps, imported the photos, and identified the species captured in the images. This protocol is the exact same process that I and my colleagues go through for setting up our own research camera traps. There are citizen science programs all over the world that let you study animals, and in some cases, you can even get close to them. For example, I've read about citizen science projects that allow you to help scientists track GPS or radio-tagged animals in the field and even observe the live capture process of administering the tracker on the animal.

A great organization that houses a lot of citizen science field expeditions all over the world is the EarthWatch Institute. You do have to pay for this program, but it's meant to replace a vacation; you get to participate in some amazing research projects. As many scientists have international field sites, they can create an EarthWatch program based on their research where people from all over the world help them collect data. EarthWatch expeditions can take you from safaris in Africa studying rhinos to investigating climate change in Antarctica. Many of these expeditions include things that I have never done before and would love to do.

The citizen science program I visited in Tsavo East National Park in Kenya was an EarthWatch program. I met two volunteers who had participated in different EarthWatch programs around the world for years. One of their favorites was a research project on sea turtles in Mexico. Because the lead scientists were permitted to handle these threatened species, the volunteers were also able to get up close with the sea turtles. All species of sea turtles are threatened or endangered, and this is not something one could normally do in ecotourism.

Quite honestly, these are some fantastic alternatives to a career in wildlife biology. Remember, once you are out of graduate school too, you are spending less and less time in the field. Two weeks on an EarthWatch field trip may be the same amount of time some professors spend in the field annually. As a volunteer, you do not have the stress of handling logistics, research equipment, managing staff, and planning budgets.

Although I don't want to be negative and discouraging, unfortunately, you do need to think about finances and job prospects when going into this career. I was personally shocked at how difficult it was for me to find a permanent job in this field in Raleigh, NC. While I recognize that I limited myself to a region, given the size of Raleigh and the fact that it, Durham, and Chapel Hill make up the "research triangle" of the United States, I thought I would be able to land a position without too much difficulty.

# GETTING A JOB IN WILDLIFE BIOLOGY

When I was in college, I chose biology as a backup major to theater so I could transition to a stable career choice if I couldn't make it as an actress. Honestly, the risk of acting scared me a lot, jumping from one gig to another without any guarantees. What would I do if I couldn't find work in between paid jobs? I wouldn't be able to find a regular job so I would have to do temp work like waiting tables, which is actually incredibly competitive in places like New York City where the acting jobs are.

Although I chose biology initially to be a medical doctor, I never thought in a million years that I would be comparing any field within biology to acting. My dad always taught me to work with my head, not with my hands. When I made the switch to wildlife biology, I felt such a sense of relief knowing I would be in a stable career. However, this is not how reality played out.

Wildlife biology seems to have become so competitive that it doesn't seem that much different from acting (except that you usually don't get to be a celebrity in wildlife biology). Many people bounce around between temporary field jobs or postdocs for years after they graduate until they secure a permanent position. I know people who have left the field completely because they couldn't get jobs.

I went into wildlife biology because I loved it; I was following my passion. And because I was happy in my career, I accepted that I wouldn't make as much as in other careers. Still, I expected to have a starting salary of least USD $60,000 with a Ph.D. Instead, I found the starting salary for Ph.D. level jobs around here (Raleigh, NC) to be in the $50,000-$55,000 range. It's not necessarily that these salaries are low but given all of your years lost to graduate school, it's a low salary to start with.

Your first salary sets the precedent for the rest of your career. In other words, it's hard to go from $50K to $80K when switching jobs and negotiating a salary, but much easier to go from $75K to $80K. Therefore, if you start with a lower salary, essentially you will be playing "catch up" for the rest of your career.

Meanwhile, my internships and graduate school experiences cost me approximately ten years of my life when I could have been making a larger salary in a different career. After graduate school, my postdoc paid me little comparatively, $40,000 per year for five years. And I was one of the lucky ones as I had no student loan debt. Many people experience a similar scenario but have tens of thousands of undergraduate student loan debt to pay on top of it. Even though I was following my passion in my career, it eventually became frustrating to make little despite working incredibly hard and caring so much about my research.

At one point, my husband was laid off from his job, leaving us to depend on my $40,000/year salary and receive support from my family. After this experience, I realized that I needed to take my salary more seriously. For most people, you have to decide if making little money or going into debt is worth it for this career.

You also have to determine if there is a market for the career that you want. You can only do this by looking at the job boards over and over again right now. For example, someone emailed me about becoming a naturalist. To my knowledge, that job doesn't exist anymore. People always tell you to follow your passion, but you need to eat.

If you've been using the job tracker this entire time, now is a good time to stop, reflect, and look at the jobs you've collected. Do they represent what you really want to do on a daily basis? Do they pay enough? How frequently do you see them or similar jobs on the job boards? If the answer is not that much, then you are going to have to become one of the best in your field, develop a strong network, or potentially come up with a backup plan should you not be able to get that kind of job.

The good news is, that after reading this book, you will have a better idea on how to be competitive in this field. The key is knowing what kind of job you want and making sure you get the experiences and skills to get you there. This field is competitive, but it will be so much easier for you because you will have this knowledge up front to make yourself a strong candidate. Additionally, I am creating new resources including courses and coaching to help guide you even further in this journey to make sure you get everything you need. You can visit my website, fancyscientist.com, to subscribe to my newsletter and be in the loop for upcoming launches.

Yes there are challenges in this career, but there are many aspects of wildlife biology that are truly wonderful: you get to ponder unanswered questions in the world, your job is meaningful and important, and you do get to travel and do things that most people would never have the opportunity to do.

Some people ask me if I regret getting a Ph.D. or wish I would have done things differently. My answer is complicated. If I had known what I know now, I would have approached getting a career in wildlife biology in a different way. I would have chosen a research project and a degree (probably a master's) that would have matched more closely to positions in the government or nonprofits and research in the United States. But the truth is, I do not regret my decision at all. All of my decisions brought me to exactly here, which is where I believe I belong.

Although I couldn't see it at the time, I am now grateful that I didn't get all of those jobs I applied for, even the ones I really wanted. Crazy, right? But this path led to me writing this book, helping you, and the launch of my career in science communication and coaching, what I've found to be my true passion.

As someone who entered this field for conservation, I've learned that reaching people is truly the biggest way I can make an impact. I first learned that people were the solution to conservation problems in 2003 when I first ventured to Kenya as a naïve young woman, but it took me another 15 years to really understand it. All of those rejections led me here.

I am so excited about my new career in science communication through my new company, the Fancy Scientist, LLC. When I think back on all of my experiences, from running from elephants in Gabon to hiking to the top of Mount Kenya, I would not have given up those once-in-a-lifetime opportunities. I think about how my acting classes and my minor in theater help me out now when I am giving public talks or creating YouTube videos to inspire people to care about wildlife. Even my own family's experience as a small business owner is now coming into play as I form my LLC. I used to not believe that things happen for a reason, but looking back on my life in retrospect, it has all prepared me for this career.

So here I am, carving my own path trying to figure it all out for my new entrepreneurial career. I hope this book provided you with the knowledge and resources to help you carve your path too. Personally, I know I can't wait to continue to write more books for you going in-depth about my experiences in Kenya and Gabon, especially to share with the world what amazing places they are and the captivating wildlife they hold. Once I figure out this whole owning-my-own business career, I'll probably write a book about that too.

# ACKNOWLEDGEMENTS

I owe a big thank you to Chris Cloney for inspiring me to turn what was supposed to be an eBook into a full-blown book. Thank you to Pat Flynn and his podcast, *Smart Passive Income*, for explaining how easy it was to self-publish and the advantages of self-publishing over traditional means. I thank my beta readers Carol Manka, Paul Manka, Lauren Pharr, Colleen Andrews, and Julia Geschke, and Kaira Wagoner for the awesome elephant photo. I thank Chana Wizenberg for encouraging me to add more stories in the beginning, allowing me to be extra vulnerable and open up about what my experiences in wildlife biology have been like.

# ABOUT THE AUTHOR

Dr. Stephanie Schuttler is a wildlife biologist, entrepreneur, and science communicator. She has 20+ peer-reviewed publications and her research has been featured in *The Washington Post*. Her studies have taken her across the globe, to all continents except for Antarctica, to study a variety of species from flour beetles to forest elephants. In 2020, she founded the Fancy Scientist, LLC. and runs the Fancy Scientist blog where she helps aspiring wildlife biologists and educates people on the science and conservation of wildlife. She believes in breaking stereotypes about what scientists look like by flaunting her love of fashion, makeup, and sparkles. Stephanie loves public speaking and has given 100+ presentations on stages across five continents. In 2019, she was invited to give a TEDx talk and is a regular contributor to the Science Channel's series "*What on Earth*." She lives in Raleigh, North Carolina with her husband, two dogs, and four cats.

Made in the USA
Columbia, SC
01 August 2024